Communications in Computer and Information Science   **2200**

AF148782

Rationale

The CCIS series is devoted to the publication of proceedings of computer science conferences. Its aim is to efficiently disseminate original research results in informatics in printed and electronic form. While the focus is on publication of peer-reviewed full papers presenting mature work, inclusion of reviewed short papers reporting on work in progress is welcome, too. Besides globally relevant meetings with internationally representative program committees guaranteeing a strict peer-reviewing and paper selection process, conferences run by societies or of high regional or national relevance are also considered for publication.

Topics

The topical scope of CCIS spans the entire spectrum of informatics ranging from foundational topics in the theory of computing to information and communications science and technology and a broad variety of interdisciplinary application fields.

Information for Volume Editors and Authors

Publication in CCIS is free of charge. No royalties are paid, however, we offer registered conference participants temporary free access to the online version of the conference proceedings on SpringerLink (http://link.springer.com) by means of an http referrer from the conference website and/or a number of complimentary printed copies, as specified in the official acceptance email of the event.

CCIS proceedings can be published in time for distribution at conferences or as post-proceedings, and delivered in the form of printed books and/or electronically as USBs and/or e-content licenses for accessing proceedings at SpringerLink. Furthermore, CCIS proceedings are included in the CCIS electronic book series hosted in the SpringerLink digital library at http://link.springer.com/bookseries/7899. Conferences publishing in CCIS are allowed to use Online Conference Service (OCS) for managing the whole proceedings lifecycle (from submission and reviewing to preparing for publication) free of charge.

Publication process

The language of publication is exclusively English. Authors publishing in CCIS have to sign the Springer CCIS copyright transfer form, however, they are free to use their material published in CCIS for substantially changed, more elaborate subsequent publications elsewhere. For the preparation of the camera-ready papers/files, authors have to strictly adhere to the Springer CCIS Authors' Instructions and are strongly encouraged to use the CCIS LaTeX style files or templates.

Abstracting/Indexing

CCIS is abstracted/indexed in DBLP, Google Scholar, EI-Compendex, Mathematical Reviews, SCImago, Scopus. CCIS volumes are also submitted for the inclusion in ISI Proceedings.

How to start

To start the evaluation of your proposal for inclusion in the CCIS series, please send an e-mail to ccis@springer.com.

Xiao-Hua Zhou · Jinzhu Jia
Editors

Causal Inference

6th Pacific Causal Inference Conference, PCIC 2024
Shanghai, China, July 5–6, 2024
Revised Selected Papers

 Springer

Editors
Xiao-Hua Zhou 🆔
Peking University
Beijing, China

Jinzhu Jia 🆔
Peking University
Beijing, China

ISSN 1865-0929 ISSN 1865-0937 (electronic)
Communications in Computer and Information Science
ISBN 978-981-97-7811-9 ISBN 978-981-97-7812-6 (eBook)
https://doi.org/10.1007/978-981-97-7812-6

This Springer imprint is published by the registered company Springer Nature Singapore Pte Ltd.
The registered company address is: 152 Beach Road, #21-01/04 Gateway East, Singapore 189721, Singapore

If disposing of this product, please recycle the paper.

Preface

We are pleased to present the proceedings of the 6th Pacific Causal Inference Conference (PCIC 2024), which was successfully held in Shanghai, China on July 5–6, 2024. The conference aimed to promote research and developmental activities in the fields of Causal Inference and Artificial Intelligence.

The conference lasted two days and included presentations by experts from all over the world. Thanks to Judea Pearl (University of California, Los Angeles, USA), Donald B. Rubin (Harvard University, USA), Xiao-Hua Zhou (Peking University, China), and Theis Lange (University of Copenhagen, Denmark). Hereby, we would like to express our sincere gratitude to all the keynote speakers and other speakers.

PCIC 2024 received a total of 15 submissions, and 8 papers were finally included. The authors in the proceedings of PCIC 2024 come from all over the world, including China, the UK, Germany, the US, and other countries. Every submitted paper underwent rigorous peer review by the conference's Scientific Committee. We sincerely thank all members for their hard work in making the conference a great success.

We extend our sincere appreciation to all the participants, presenters, organizers, and supporters who contributed to the success of PCIC 2024. Their dedication and enthusiasm were instrumental in making this conference a valuable and enriching experience.

It is our hope that the proceedings of PCIC 2024 will serve as a lasting resource, capturing the knowledge and ideas shared during the conference and inspiring further progress in the field of Causal Inference and Artificial Intelligence.

We look forward to future opportunities for continued collaboration and knowledge sharing within the global community of Causal Inference and Artificial Intelligence.

October 2024   Jinzhu Jia

Organization

Scientific Committee Chair

Xiao-Hua Zhou — Peking University, China

Local Organizing Committee Chairs

Riquan Zhang — Shanghai University of International Business and Economics, China

Yong Zhou — East China Normal University, China

Scientific Committee Members

Ruichu Cai	Guangdong University of Technology, China
Peng Ding	University of California, Berkeley, USA
Fang Han	University of Washington, USA
Satoshi Hattori	Osaka University, Japan
Jinzhu Jia	Peking University, China
Kajsa Kvist	Novo Nordisk, Denmark
Theis Lange	University of Copenhagen, Denmark
Fabrizia Mealli	European University Institute, Italy
Wang Miao	Peking University, China
Yumou Qiu	Peking University, China
Thomas S. Richardson	University of Washington, USA
Manuel Gomez Rodriguez	Max Planck Institute for Software Systems, Germany
Ricardo Silva	University College London, UK
Lan Wang	University of Miami, USA
Linbo Wang	University of Toronto, Canada
Lu Wang	University of Michigan, USA
Shu Yang	North Carolina State University, USA
Ting Ye	University of Washington, USA
Kun Zhang	CMU, USA & MBZUAI, UAE

Local Organizing Committee Members

Jie Chen	Shanghai University of International Business and Economics, China
Rui Li	Shanghai University of International Business and Economics, China
Luyao Wang	East China Normal University, China
Yi Wang	Shanghai University of International Business and Economics, China
Ying Zhang	East China Normal University, China

Contents

Avoiding the Unconfoundednes Assumption: Counterfactual Inference
Considering Unobserved Confounders 1
  Yonghe Zhao, Yun Peng, and Huiyan Sun

Causal Inference in the Multiverse of Hazard 15
  En-Yu Lai and Yen-Tsung Huang

A Continuous Structural Intervention Distance to Compare Causal Graphs 25
  *Mihir Dhanakshirur, Felix Laumann, Junhyung Park,
  and Mauricio Barahona*

Detection Windows from Hidden Markov Model for Discovering Varying
Causal Relations Between Time Series 41
  Kaijun Wang, Ying Fang, and Tianjian Luo

Real-World Implications of a Methodological Dilemma: Endogenous
Confounding in Causal Decomposition Analysis 49
  Ha-Joon Chung and Guanglei Hong

Evaluation Criteria for Causal Discovery Without Ground-Truth Graphs 65
  Lei Wang, Shanshan Huang, Liao Jun, and Li Liu

Optimizing Experimental Design for Causal Effect Estimation with Partial
Measurements 74
  Leopold Mareis

Exploring the Use of Q-Learning in Causal Inference for Adaptive
Interventions 86
  *Sha Zhou, YanHua Jiang, ZhiWei Jin, ZhenZhen Qian, MengMeng Ji,
  Chi Liu, HongYi Li, GuoWei Xuan, YuXing Shuai, and XinLin Chen*

Author Index 95

Avoiding the Unconfoundednes Assumption: Counterfactual Inference Considering Unobserved Confounders

Yonghe Zhao[1]ⓘ, Yun Peng[2]ⓘ, and Huiyan Sun[1](✉)ⓘ

[1] School of Artificial Intelligence, Jilin University, Changchun, China
yhzhao21@mails.jlu.edu.cn, huiyansun@jlu.edu.cn
[2] Department of Data Analysis, Baidu Netdisk, Beijing, China

Abstract. Counterfactuals are the basis of causal inference in observational studies. The fundamental challenge in counterfactual inference lies in addressing the impact of both observed and unobserved confounders. Despite the multitude of proposed methods aimed at addressing observed confounding bias, these approaches are contingent upon the untestable assumption of Unconfoundedness, which posits the absence of unobserved confounders. In this paper, we present a practical framework of Counterfactual Inference Considering Unobserved Confounders (CIUC), which effectively addresses the influence of both observed and unobserved confounders, thereby enabling precise estimation of counterfactual outcomes. Specifically, the CIUC framework begin by employing variational learning to derive the distribution of unobserved confounders disentangled from observed covariates, relaxing the untestable assumption of Unconfoundedness. Furthermore, within the framework, we incorporate a balanced representation model that takes into account both observed and unobserved confounders. This integration guarantees the attainment of unbiased inferences. The CIUC framework is versatile, as it can accommodate both discrete and continuous treatment variables. Additionally, it seamlessly integrates with various existing counterfactual inference models, making it applicable across a wide range of scenarios. In contrast to the majority of existing methods, the CIUC framework goes beyond by offering confidence intervals for the counterfactual outcomes, which proves highly advantageous for risk-sensitive tasks. Extensive experiments conducted on synthetic, semi-synthetic, and real-world datasets provide compelling evidence of the CIUC's outstanding performance in generating unobserved confounders, learning balanced representations, and accurately estimating treatment effects at both group and individual levels.

Keywords: Counterfactual Inference · Unobserved Confounder · Confounding Adjustment · Variational Learning · Balanced Representation

ⓒ The Author(s), under exclusive license to Springer Nature Singapore Pte Ltd. 2025
X.-H. Zhou and J. Jia (Eds.): PCIC 2024, CCIS 2200, pp. 1–14, 2025.
https://doi.org/10.1007/978-981-97-7812-6_1

1   Introduction

Causality is widely known to represent a more fundamental relationship between variables, revealing the directionality and determinacy [13]. Therefore, causal inference has attracted increased attention in various domains, such as epidemiology, healthcare, and economics [1,11,16], etc. Randomized controlled trials (RCTs) are a reliable method for investigating causal relationships between variables. However, RCTs can be time-consuming, expensive, and may involve ethical considerations in certain scenarios [7,21]. How to estimate causality from observational data is the critical issue and draws wide attention.

In contrast to the randomized treatment assignment in RCTs, the primary challenge in causal inference for observational studies lies in the unknown mechanism of treatment assignment [13]. In other words, observational data often exhibit various deviations, such as the presence of confounders that influence both the treatment and outcome variables. The presence of confounding bias can result in unsatisfactory accuracy of inference when utilizing an inferential model, denoted as $m_A(x)$, which is fitted using data from group A, to estimate the effect of treatment B. This scenario is analogous to the domain adaptation problem [15].

Given the inherent limitations of observational data, confounders are categorized as either observed or unobserved. Numerous inference methods have been developed within the Potential Outcome Framework (POF) to address the observed confounders. The typical methods include Re-weighting [25], Stratification [5], Matching [4], tree model based [6], representation learning based [2,15,16] and meta learning based models [18,24], etc. However, relatively few approaches have been developed to mitigate the influence of unobserved confounders. While methods such as CEVAE [20] focus on identifying latent confounding variables by leveraging observed covariates as proxies, the issue of unobserved confounders without suitable proxies also deserves significant attention. Furthermore, VLUCI [32] introduced a variational learning framework to infer unobserved confounders, which does not adequately specify the procedures for conducting deconfounding inference. Most other methods commonly rely on the assumption of Unconfoundedness to circumvent the discussion and potential misinformation caused by unobserved confounders.

In this paper, our focus is on addressing both observed and unobserved confounders. To tackle this challenge, we propose a practical framework for counterfactual inference called CIUC. Specifically, the proposed CIUC framework consists of three networks: Factual Prediction Network (FPN), Variational Learning Network (VLN), and Counterfactual Inference Network (CIN). Firstly, the FPN is designed to quantify the factual mappings of observed covariates to treatment and outcome variables, and further to exclude the influence of covariates for treatment and outcome variables individually. Secondly, an interactive doubly variational network, VLN, is applied to infer the distribution of unobserved confounders based on the results acquired from the FPN. Finally, the balanced representation model CIN is constructed to infer unbiased counterfactual outcomes by considering both the observed covariates and the unobserved confounders. The main contributions of this paper are as follows.

(1) We develop a practical CIUC framework that effectively addresses the impact of both observed and unobserved confounders, resulting in precise counterfactual inference. By relaxing the untestable assumption of Unconfoundedness, the framework significantly enhances the practicability of causal inference in observational studies, thereby enabling more robust and insightful analyses in real-world settings.

(2) The CIUC framework is adaptive to discrete or continuous treatment variables and is compatible with various existing counterfactual models for confounding adjustment. Moreover, leveraging the inferred distribution of unobserved confounders, CIUC framework offers interval estimation of counterfactual outcomes that provides more favorable support for the risk-sensitive decision.

(3) Extensive experiments on synthetic, semi-synthetic, and real-world datasets demonstrate that the CIUC framework significantly outperforms various state-of-the-art counterfactual inference models on the precision and stability of causal inference.

2   Related Works

2.1   Statistics-Based Methods

Within the POF, various statistics-based models, including Re-weighting [25], Stratification [5], and Matching [4], are employed to mitigate the impact of confounding bias. Specifically, the Re-weighting method employs inverse propensity score weights to estimate average treatment effects (ATE). Robins et al. introduced a doubly robust estimator [23] that extends Re-weighting to the assessment of individual treatment effects (ITE) by incorporating linear regression. In addition, Stratification assesses subgroup causal effects by dividing the sample based on confounders. Matching, on the other hand, derives counterfactual estimates by identifying the nearest neighbors in the control group to the treatment group. These statistics-based methods assume linear causality and the availability of appropriate nearest neighbors, conditions that are not always fulfilled in practical scenarios.

2.2   Machine Learning-Based Methods

High-dimensional confounders challenge the traditional statistics-based methods. Machine learning's predictive strength has spurred several counterfactual inference algorithms. Chipman et al. introduced the Bayesian Additive Regression Tree (BART) [6]. This ensemble algorithm leverages the inherent hierarchical structure of decision trees to address confounding bias. In addition, representation-based learning algorithms, TarNet [26], CFR [26] and SITE [30], utilize the capability of neural networks to obtain balanced representation of confounders. Further, the GANITE [31], leveraging generative adversarial networks, infers counterfactual outcomes and assesses ITE. These machine learning-based methods presuppose Unconfoundedness assumption, which is unverifiable. To address this challenge, the generative model CEVAE [20] utilizes proxy variables to generate posterior distributions of latent confounders. The CEVAE model softens the Unconfoundedness assumption, but it relies on the availability of appropriate proxy variables for latent confounders. However, it is crucial to recognize that the absence of suitable proxies for unobserved confounders remains a significant concern that requires careful consideration.

3   Method

3.1   Problem Setting

The CIUC framework is based on the POF and primarily aims at observational data, represented as $\mathcal{D} = \{X_i, U_i, t_i, y_i\}_{i=1}^{N}$, which encompasses N independent and identically distributed samples. Referring to the causal structure shown in Fig. 1, not only the treatment variable t and the covariates X are causally related to the outcome variable y, but also the unobserved confounders U. In practice, only the factual outcome $y_i^f(t_i)$ corresponding to t_i is observable, while the other counterfactual outcomes $y_i^{cf}(C_T t_i)$ are not accessible. The primary task of the CIUC framework is counterfactual inference for $y_i^{cf}(C_T t_i)$. For generalization, the POF establishes the Unconfoundedness assumption, implying the absence of unobserved confounders U. While the CIUC framework relaxes this unverifiable assumption, indicating the existence of U, i.e., $t \perp y|X, U$. Moreover, the validity of the CIUC framework relies on two other essential assumptions in POF [29]: Stable Unit Treatment Value and the Positivity assumption.

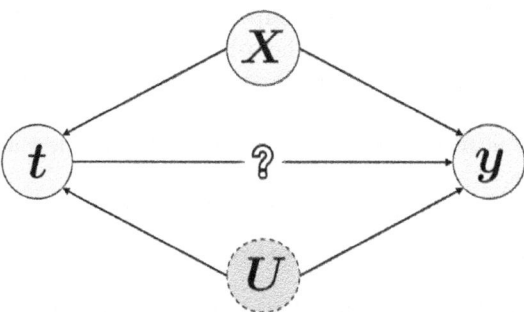

Fig. 1. The causal structure diagram within problem setting. Where yellow squares represent observed variables, otherwise unobserved. (Color figure online)

3.2   The Practical Framework for Counterfactual Inference Considering Unobserved Confounders

The Overview of the CIUC Framework. The overview of the CIUC framework is depicted in Fig. 2, comprising three sub-networks, including FPN, VLN, and CIN. Upon examining Fig. 1, it becomes apparent that there is no confounder between the observed covariates X and both the treatment t and outcome y. Consequently, drawing upon the principles of identifiability theory [32], we develop the FPN, as illustrated in Fig. 2a, which enables the unbiased estimation of the causal effects of X on t and y. Then, based on the exclusion of segments of the t and y that were influenced by the X, we adopt the VLN shown in Fig. 2b, thus allowing for the inference of the distribution of unobserved confounders U. In the final stage, exemplified by binary treatment, we amalgamate the observed X with the identified U to propose the CIN based on balanced representation learning, as shown in Fig. 2c. The CIUC framework progresses through a cascade training approach across each network, each of which mirrors the underlying causal structure. The following section will elaborate on the intricacies of each sub-network.

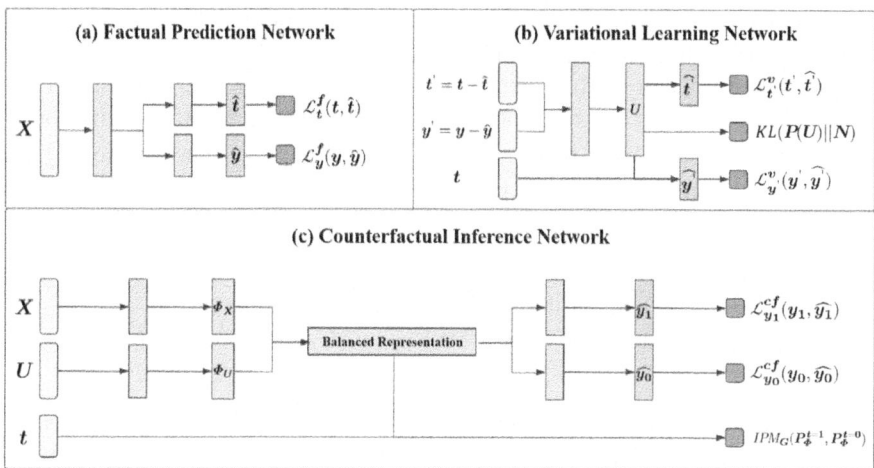

Fig. 2. The overview of the practical framework CIUC that contains three sub-networks: (a) Factual Prediction Network (FPN); (b) Variational Learning Network (VLN); (c) Counterfactual Inference Network (CIN).

Factual Prediction Network. From the causal structure depicted in Fig. 1, it is evident that the covariates X are exogenous variables, which makes the causal effects of X on the treatment t and outcome y identifiable [32]. Consequently, the construction of the FPN, $\hat{t} = f(X; W_X^t)$ and $\hat{y} = f(X; W_X^y)$, enables the acquisition of unbiased estimates of the effects of X on t and y, respectively. The FPN performs the role of feature selection, that is, the filtering of the X that affect t and y during the training process. In addition, the FPN prepares for inferring unobserved confounders independent of the covariates. As shown in Eq. (1), the objective function of FPN includes the prediction errors ($\mathcal{L}_t^f$ and $\mathcal{L}_y^f$) and the penalty term $\lambda^f \|W\|^2$ which avoids overfitting. In addition, the prediction error of the continuous treatment variables is suggested to replace the cross entropy (CE) with the mean square error (MSE).

$$\mathcal{L}^f = \mathcal{L}_t^f + \mathcal{L}_y^f + \lambda^f \|W\|^2$$

$$\text{Where,} \begin{cases} \mathcal{L}_t^f = \text{CE}(t, \hat{t}) = -\frac{1}{n} \sum_{i=1}^{n} [t_i log(\hat{t}_i) + (1 - t_i) log(1 - \hat{t}_i)] \\ \mathcal{L}_y^f = \text{MSE}(y, \hat{y}) = \frac{1}{n} \sum_{i=1}^{n} (y_i - \hat{y}_i)^2 \end{cases} \quad (1)$$

Variational Learning Network. According to Fig. 1, the unobserved confounders U is an exogenous variable and is independent of the X. Therefore, based on the FPN, we obtained $t' = t - \hat{t}$ and $y' = y - \hat{y}$ that are disentangled with the X. Further, we assume that U obeys a multivariate Gaussian distribution: $P(U) \sim N(0, I)$, where the dimension of U is a hyperparameter [9]. Since U is a causal variable of t' and y', with t' and y' as inputs, the VLN is utilized to infer the distribution of unobserved confounders $P(U|t', y')$ in a generative method. The VLN is a doubly variational generative networks, whose objective function $\mathcal{L}^v$ consists of distribution errors $\mathcal{L}_{P(U)}^v$ represented by the Kullback-Leibler(KL) divergence and reconstruction errors $\mathcal{L}_{t',y'}^v$, as shown in Eq. (2); for detailed derivation, refer to [17,32].

$$\mathcal{L}^v = \mathcal{L}_{P(U)}^v + \mathcal{L}_{t',y'}^v$$

$$\text{Where,} \begin{cases} \mathcal{L}_{P(U)}^v = \text{KL}(P(U|t', y') \| N(0, I))) \\ \mathcal{L}_{t',y'}^v = \text{MSE}(t', \hat{t'}(\hat{U})) + \text{MSE}(y', \hat{y'}(\hat{U}, t)), \ \hat{U} \sim P(U|t', y') \end{cases} \quad (2)$$

Counterfactual Inference Network. After the training of FPN and VLN, we acquired the distribution of unobserved confounders. Furthermore, the primary concern in CIN is the confounding bias. With binary treatment as an example, we improve the multi-headed CFR [26] model by considering the learned unobserved confounders by VLN, as shown in Fig. 2c. To avoid the problem that the difference of the dimension between covariates and unobserved confounders leads to one side not being optimized, we provide two parallel balanced representation learning in CIN. An important aspect to highlight is the flexibility of the CIN in handling both discrete and continuous treatment variables. This adaptability allows the CIN to seamlessly integrate with state-of-the-art deconfounding methods that are specifically designed for different treatment types, as described above. The CIN is constructed based on the principle of deconfounding with balanced representation, where the balanced criterion is adopted from the Integral

Probability Metrics (IPM) [22,27], as shown in Eq. (3).

$$\text{IPM}_G(P_\Phi^{t=0}, P_\Phi^{t=1}) = \sup_{g \in G} \left| \int_S g(s)(P_\Phi^{t=0}(s) - P_\Phi^{t=1}(s))ds \right| \qquad (3)$$

where $P_\Phi^{t=0}$ and $P_\Phi^{t=1}$ denote the distribution on confounding representations of control and treatment groups.

This paper use a specific IPM: the Maximum Mean Discrepancy [10]. Shalit et al. [26] demonstrated that the expected errors for the CIN is the sum of the discrepancy of the confounding representation under the IPM criterion and the standard generalization error $\mathcal{L}_y^{cf}$ shown in Eq. (4). Further, to improve the generalization performance of the network, the objective function $\mathcal{L}^c$ of the CIN with the additional penalty term is shown in Eq. (5).

$$\mathcal{L}_y^{cf} = \text{MSE}(y, f(t, \Phi_X, \Phi_U)) = \frac{1}{n} \sum_{i=1}^{n} (y_i - f(t_i, \Phi_{X_i}, \Phi_{U_i}))^2 \qquad (4)$$

$$\mathcal{L}^c = \mathcal{L}_y^{cf} + \text{IPM}_G(P_{\Phi_X}^{t=0}, P_{\Phi_X}^{t=1}) + \text{IPM}_G(P_{\Phi_U}^{t=0}, P_{\Phi_U}^{t=1}) + \lambda^c \|W\|^2 \qquad (5)$$

4   Experiments

4.1   Datasets

In practice, researchers often encounter a situation where the counterfactual outcomes in the real-world are unobservable. In addition, the distribution of unobserved confounders is not available. To address these limitations, we use three types of datasets, including a synthetic dataset, the semi-synthetic dataset: Infant Health and Development Program (IHDP), and the real-world dataset: Jobs, to compare the performance of the proposed framework with various state-of-the-art methods.

Synthetic Dataset. Based on the generation schema in Fig. 1, the synthetic dataset $\mathcal{D}^S$ consisted of four components: $\{t, X, U, y\}$. Firstly U and X are randomly generated as shown in Eq. (6), which are exogenous variables. In the experiments, U as ground-truth is not used for training. The inferential performance of the proposed framework is verified by comparing the distribution of U learned by the VLN with the ground-truth.

$$U \sim N(0,1); \quad X \sim N(0^{1 \times d}, \frac{1}{2}(\Sigma + \Sigma^T)), \Sigma \sim \text{Unif}((-10,10)^{d \times d}) \qquad (6)$$

Next, t and y are generated by the non-linear equation shown in Eq. (7)-(8).

$$P(t) \sim \mathcal{B}(\lambda_X * f(X \cdot W_{Xt} + b_{Xt}) + \lambda_U * f(U \cdot W_{Ut} + b_{Ut}) + \lambda_\epsilon * \epsilon_t) \qquad (7)$$

$$y = f(X \cdot W_{Xy} + b_{Xy}) + f(U \cdot W_{Uy} + b_{Uy}) + (t \cdot W_{ty} + b_{ty})^2 + \epsilon_y \qquad (8)$$

where, function f = sigmoid, parameters $W_X. \sim \text{Unif}(-5,5)$, $W_U. = 3$, $W_{ty} = 2$, $b \sim N(0,0.1)$, $\epsilon_t \sim \text{Unif}(0,1)$, and $\epsilon_y \sim N(0,0.1)$. As shown in Eq. (7), t obeys a Bernoulli distribution $\mathcal{B}(X, U, \epsilon_t)$, where $\lambda.$ is the weighting coefficient, $\lambda_X + \lambda_U + \lambda_\epsilon = 1$.

Semi-synthetic Dataset: IHDP. The IHDP dataset is a semi-synthetic dataset based on a RCT in the Infant Health and Development Program [12]. The dataset comprises 747 units(139 treated, 608 control) and 25 covariates measuring aspects of children and their mothers. The treatment variable is home visits with specialists, and the outcome variable is children's cognitive test scores. To obtain biased observational data, the treated assignment was then 'de-randomized' by removing children whose mother is non-white from the treatment group. As a result, the treatment and control groups are no longer balanced and simple comparisons of outcomes lead to biased estimates of treatment effects. We follow the setting in [26] and use 100 replications of the simulated outcomes.

Real-World Dataset: Jobs. The Jobs dataset is a widely used benchmark dataset in the field of causal inference regarding the studies of job training [19]. A detailed statistical description of this dataset is given in [8]. In brief, the Jobs dataset includes eight covariates, including age, education, black, Hispanic, married, no-degree, and pre-training earnings. The treatment variable is job training and the outcome variable is post-training unemployment. Following [26], the Jobs dataset includes Lalonde experimental sample(297 treated, 425 control) and the PSID comparison group(2490 control).

4.2   Evaluation Index

To illustrate the performance of the proposed CIUC framework, the errors of ITE and ATE are used to assess inferential accuracy at the individual and group levels, respectively, as shown in Eq. (9).

$$\text{ITE}_i = y^i(1) - y^i(0); \quad \text{ATE} = \mathbb{E}[y(1) - y(0)] \tag{9}$$

Specifically, for synthetic and semi-synthetic datasets in which the counterfactual outcomes are accessible, the precision in the estimation of heterogeneous effects (PEHE) [12] is an evaluation metric for the performance of estimations on ITE, as shown in Eq. (10).

$$\epsilon_{\text{PEHE}} = \frac{1}{N} \sum_{i=1}^{N} ([y^i(1) - y^i(0)] - [\hat{y}^i(1) - \hat{y}^i(0)])^2 \tag{10}$$

where y^i represents the ground-truth outcome, and $y^i(1) - y^i(0)$ reflects the actual value of the ITE. On the other hand, $\hat{y}^i$ denotes the predicted outcome, and $\hat{y}^i(1) - \hat{y}^i(0)$ represents the predicted value of the ITE.

In addition, ϵ_{ATE}, the absolute error of ATE is used to measure model performance at the group level, as shown in Eq. (11).

$$\epsilon_{\text{ATE}} = |\text{ATE} - \hat{\text{ATE}}| \tag{11}$$

For the real-world dataset, Jobs, counterfactual outcomes and ground-truth ITE are not observable. Therefore, the PEHE metric is no longer applicable. The policy risk [26] shown in Eq. (12) is utilized to measure errors on ITE.

$$
\begin{aligned}
R_{\text{Pol}}(\pi_f) = 1 &- \mathbb{E}[y(1)|\pi_f(X) = 1] \cdot P(\pi_f = 1) \\
&- \mathbb{E}[y(0)|\pi_f(X) = 0] \cdot P(\pi_f = 0)
\end{aligned}
\tag{12}
$$

where, $\pi_f(X) = 1$ represents that there is a difference between the inferred treated and the inferred control effect of the model, that is $f(X, t = 1) - f(X, t = 0) > \lambda$, otherwise $\pi_f(X) = 0$.

In addition, considering that the Jobs dataset contains a subset of RCT, we acquired the average treatment effect on the treated group (ATT) as a benchmarking effect and the absolute error ϵ_{ATT} of ATT calculated in Eq. (13) [29].

$$
\text{ATT} = \frac{1}{N_T} \sum_{i \in T} y^i - \frac{1}{N_C} \sum_{i \in C} y^i; \quad \epsilon_{\text{ATT}} = \left| \text{ATT} - \frac{1}{N_T} \sum_{i \in T} (\hat{y}^i(1) - \hat{y}^i(0)) \right| \tag{13}
$$

where, T and C represents treated and control group of RCT subset of the Jobs dataset.

4.3 Experimental Design

In this paper, the baseline models are various state-of-the-art counterfactual inference models based on ensemble learning and neural networks: including BART [6], R Forest [3], C Forest [28], BNN [14], TarNet [26], CFRW [26], CEVAE [20], and GANITE [31].

We will validate the performance of the proposed framework on synthetic data for the tasks of generating unobserved confounders, learning balanced representation, and counterfactual inference. The two benchmark datasets IHDP and Jobs are randomly divided into training/test sets according to the percentages of 80/20. Under this division criterion, the model to be evaluated is repeated 100 times to record the mean and standard error of the evaluation metrics. We give the values in contexts including in-sample and out-sample for the above evaluation metrics. Where in-sample refers to the training set and out-sample refers to the test set.

4.4 Results

Result on the Synthetic Dataset. For the synthetic dataset, under the setting: $d = 5, \lambda_X = 0.6, \lambda_U = 0.35$, the generated distribution of unobserved confounders is provided in Fig. 3a. The results illustrate that the generated distribution by the proposed framework coincides with the unobserved confounders which are not involved in training.

For the balanced representation learning, Fig. 3c-3d shows the variations of the correlations of the treatment and covariates after the training. Figure 3c shows the violin plot and the heat map of treatment and covariates in the

$\mathcal{D}^S$, illustrating the differences in the distribution of the covariates between the treated and control group. Otherwise, Fig. 3d indicates that the correlation was significantly reduced after modeling, and the framework yields a balanced representation of the covariates.

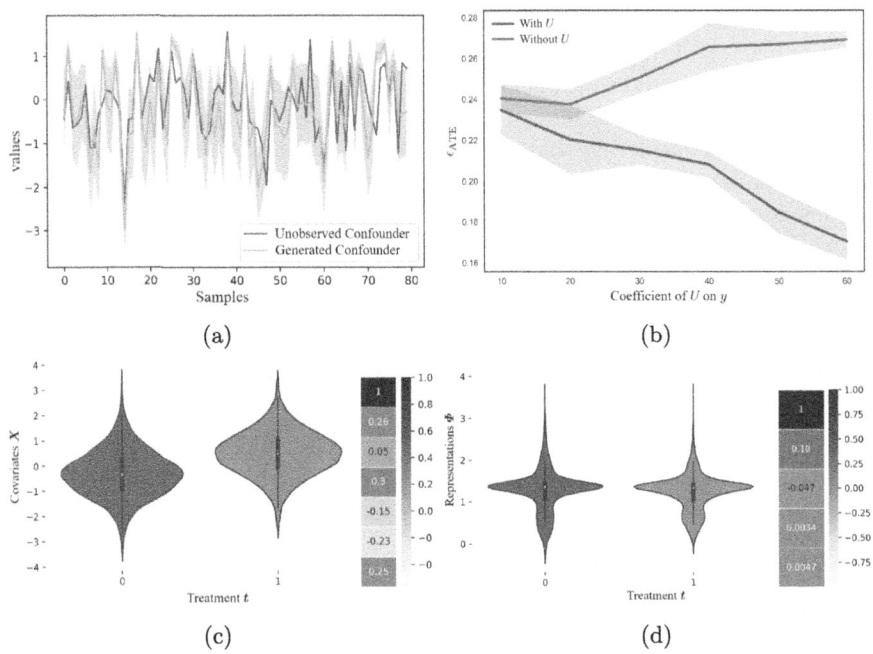

Fig. 3. The performance of CIUC on the synthetic dataset. (a) The generated distribution of unobserved confounders. (b) The inference errors in various coefficient of U on y. (c) The correlations of the treatment and covariates on the original dataset. (d) The correlations of the treatment and the generated representations through CIN.

Finally, we evaluated the performance of the CIUC framework for counterfactual inference when, W_{Uy}, the influence coefficients of the U on the y change in the range of set $\{10, 20, \cdots, 60\}$. Figure 3b shows the ϵ_{ATE} and their standard deviations on the $\mathcal{D}^S$. The red line represents the proposed framework, while the blue line represents the direct training of the counterfactual model that ignores unobserved confounders. As shown in Fig. 3b, the proposed framework improved the accuracy as the coefficients W_{Uy} increased. The phenomenon suggests that the CIUC framework remarkably improves the performance of counterfactual inference when the W_{Uy} is significant.

Result on the Benchmark Dataset. Further, the dataset IHDP with counterfactual outcomes and real-world dataset Jobs are available to validate the performance of the framework in estimating causal effects at the individual and group level, respectively, as shown in Table 1.

Table 1. The performance of ITE estimation with two benchmark datasets. Where In-Sa. refers to the training and validation set and Out-Sa. refers to the test set.

Methods	Benchmark Datasets							
	IHDP(ϵ_{ATE})		IHDP($\sqrt{\epsilon_{PEHE}}$)		Jobs(ϵ_{ATT})		Jobs(R_{Pol})	
	In-Sa.	Out-Sa.	In-Sa.	Out-Sa.	In-Sa.	Out-Sa.	In-Sa.	Out-Sa.
BART	.23 ± .01	.34 ± .02	2.1 ± .13	2.3 ± .14	.02 ± .00	.08 ± .03	.23 ± .01	.25 ± .02
R. Forest	.73 ± .05	.96 ± .06	4.2 ± .23	6.6 ± .31	.03 ± .01	.09 ± .04	.23 ± .01	.28 ± .02
C. Forest	.18 ± .01	.40 ± .03	3.8 ± .21	3.8 ± .22	.03 ± .01	.07 ± .03	.19 ± .00	.29 ± .02
BNN	.37 ± .03	.42 ± .03	2.2 ± .14	2.1 ± .14	.04 ± .01	.09 ± .04	.20 ± .01	.24 ± .02
TarNet	.26 ± .01	.28 ± .01	.88 ± .02	.95 ± .02	.05 ± .02	.11 ± .04	.17 ± .01	.21 ± .01
CFRW	.25 ± .01	.27 ± .01	.71 ± .02	.76 ± .02	.04 ± .01	.09 ± .03	.17 ± .01	.21 ± .01
CEVAE	.34 ± .01	.46 ± .02	2.7 ± .14	2.6 ± .12	**.02 ± .01**	**.03 ± .01**	**.15 ± .01**	.26 ± .02
GANITE	.43 ± .05	.49 ± .05	1.9 ± .42	2.4 ± .44	**.02 ± .01**	.06 ± .03	**.15 ± .01**	.17 ± 0.01
Ours	**.06 ± .04**	**.07 ± .06**	**.52 ± .08**	**.53 ± .09**	**.02 ± .01**	**.03 ± .01**	**.15 ± .01**	**.16 ± .06**

For the IHDP dataset, the proposed framework significantly outperformed the state-of-the-art models at both individual and population levels. While, for the Jobs dataset, the estimation performance of the framework is improved or comparable to the baseline models. In fact, the ATE on the Jobs dataset is only 0.08, which indicates that the effect of the treatment is not significant and there is little space for the model to be improved. It is worth mentioning that the proposed framework demonstrates favorable performance in generalization within and outside the sample. In addition, the nature of the multivariate Gaussian distribution of the generated U yields a confidence interval estimate for subsequent counterfactual inference, which provides more reliable and valuable information for decision-making, especially in risk-sensitive domains.

It is worth noting that the VLN generates the representation of unobserved confounders, but not the actual unobserved confounders themselves. In all three cases investigated, it consistently emerged that only a few dimensions were crucial for capturing the information of unobserved confounders. Thus, setting a high number of dimensions is unnecessary. In the application of the CIUC framework, inferring the entities behind the generated unobserved confounders in conjunction with real-world scenarios can indeed provide valuable insights into the intrinsic causal relationships between variables.

5   Conclusion

Causal inference from observational data is widely recognized and holds immense significance across various disciplines, including economics, healthcare, and recommendation systems. Handling the influence of both observed and unobserved confounders represents a primary challenge in causal inference. To overcome this challenge, we propose a practical framework called CIUC, which comprises three interconnected networks: the FPN, VLN, and CIN. The proposed CIUC framework effectively mitigates confounding bias and enables precise inference of counterfactual outcomes. To elaborate, the FPN and VLN components of the CIUC framework aim to learn the distribution of unobserved confounders disentangled with the observed confounders, thereby relaxing the unverifiable assumption of Unconfoundedness. Subsequently, the CIN acquires a balanced representation that incorporates both the observed and unobserved confounders to facilitate unbiased inference. The CIUC framework is adaptive to discrete or continuous treatment variables, which allows for its application in a wide range of scenarios. In particular, CIUC framework additionally provides the interval estimates of counterfactual outcomes for precise decisions. Extensive experiments conducted on synthetic, semi-synthetic, and real-world datasets demonstrate the superior performance of the CIUC framework compared to various state-of-the-art counterfactual inference models. In the future, we are interested in exploring the unobserved confounders in more complex causal structures and investigating their interpretability.

Acknowledgments. The authors thank funding support from the National Natural Science Foundation of China (No. 62372210), Natural Science Foundation of Jilin Province (20240101025JJ).

Disclosure of Interests. The authors have no competing interests.

References

1. Alaa, A.M., van der Schaar, M.: Bayesian inference of individualized treatment effects using multi-task Gaussian processes. In: Advances in Neural Information Processing Systems, vol. 30. Curran Associates, Inc. (2017)
2. Ben-David, S., Blitzer, J., Crammer, K., Pereira, F.: Analysis of representations for domain adaptation. Adv. Neural Inf. Process. Syst. **19** (2006)
3. Breiman, L.: Random forests. Mach. Learn. **45**, 5–32 (2001)
4. Castro-Martín, L., del Mar Rueda, M., Ferri-García, R.: Combining statistical matching and propensity score adjustment for inference from non-probability surveys. J. Comput. Appl. Math. **404**, 113414 (2022)
5. Chang, Y., Dy, J.: Informative subspace learning for counterfactual inference. In: Proceedings of the AAAI Conference on Artificial Intelligence, vol. 31 (2017)
6. Chipman, H.A., George, E.I., McCulloch, R.E.: BART: Bayesian additive regression trees. Ann. Appl. Stat. **4**(1), 266–298 (2010)
7. D'Aunno, T.: Reputation and power: organizational image and pharmaceutical regulation at the FDA. Adm. Sci. Q. **55**(4), 671–672 (2010)

8. Dehejia, R.H., Wahba, S.: Propensity score-matching methods for nonexperimental causal studies. Rev. Econ. Stat. **84**(1), 151–161 (1998)
9. Doersch, C.: Tutorial on variational autoencoders. arXiv e-prints (2016)
10. Gretton, A., Borgwardt, K.M., Rasch, M., Schlkopf, B., Smola, A.J.: A kernel two-sample test. J. Mach. Learn. Res. **13**, 723–773 (2012)
11. Hannart, A., Pearl, J., Otto, F., Naveau, P., Ghil, M.: Causal counterfactual theory for the attribution of weather and climate-related events. Bull. Am. Meteor. Soc. **97**(1), 99–110 (2016)
12. Hill, J.: Bayesian nonparametric modeling for causal inference. J. Comput. Graph. Stat. **20**, 217–240 (2011)
13. Imbens, G.W., Rubin, D.B.: Causal Inference in Statistics, Social, and Biomedical Sciences. Cambridge University Press, Cambridge (2015)
14. Johansson, F.D., Shalit, U., Sontag, D.: Learning representations for counterfactual inference. In: International Conference on Machine Learning, pp. 3020–3029 (2016)
15. Johansson, F.D., Kallus, N., Shalit, U., Sontag, D.: Learning weighted representations for generalization across designs. arXiv preprint arXiv:1802.08598 (2018)
16. Johansson, F.D., Shalit, U., Sontag, D.: Learning representations for counterfactual inference. arXiv e-prints (2016)
17. Kingma, D.P., Welling, M.: Auto-encoding variational Bayes. arXiv e-prints arXiv:1312.6114 (Dec 2013)
18. Künzel, S.R., Sekhon, J.S., Bickel, P.J., Yu, B.: Meta-learners for estimating heterogeneous treatment effects using machine learning. Proc. Natl. Acad. Sci. **116**, 4156–4165 (2017)
19. Lalonde, R.J.: Evaluating the econometric evaluations of training programs with experimental data. Working Papers (1984)
20. Louizos, C., Shalit, U., Mooij, J., Sontag, D., Zemel, R., Welling, M.: Causal effect inference with deep latent-variable models. In: Advances in Neural Information Processing Systems, vol. 30, pp. 1049–5258. Curran Associates, Inc. (2017)
21. Malina, D., Bothwell, L.E., Greene, J.A., Podolsky, S.H., Jones, D.S.: Assessing the gold standard-lessons from the history of RCTs. N. Engl. J. Med. **374**(22), 2175–2181 (2016)
22. Müller, A.: Integral probability metrics and their generating classes of functions. Adv. Appl. Probab. **29**(2), 429–443 (1997)
23. Robins, J.M., Rotnitzky, A., Zhao, L.P.: Estimation of regression coefficients when some regressors are not always observed. J. Am. Stat. Assoc. **89**(427), 846–866 (1994)
24. Robinson, P.M.: Root-n-consistent semiparametric regression. Econometrica **56**(4), 931–954 (1988)
25. Rosenbaum, P.R., Rubin, D.B.: The central role of the propensity score in observational studies for causal effects. Biometrika **70**(1), 41–55 (1983)
26. Shalit, U., Johansson, F.D., Sontag, D.: Estimating individual treatment effect: generalization bounds and algorithms. arXiv e-prints (2017)
27. Sriperumbudur, B.K., Fukumizu, K., Gretton, A., SchöLkopf, B., Lanckriet, G.: On the empirical estimation of integral probability metrics. Electron. J. Stat. **6**(6), 1550–1599 (2012)
28. Wager, S., Athey, S.: Estimation and inference of heterogeneous treatment effects using random forests. Res. Papers **8**(6), 1831–45 (2017)
29. Yao, L., Chu, Z., Li, S., Li, Y., Gao, J., Zhang, A.: A survey on causal inference. ACM Trans. Knowl. Discov. Data (TKDD) **15**(5), 1–46 (2021)

30. Yao, L., Li, S., Li, Y., Huai, M., Gao, J., Zhang, A.: Representation learning for treatment effect estimation from observational data. Adv. Neural Inf. Process. Syst. **31** (2018)
31. Yoon, J., Jordon, J., van der Schaar, M.: GANITE: estimation of individualized treatment effects using generative adversarial nets. In: International Conference on Learning Representations (2018)
32. Zhao, Y., Huang, Q., Wu, S., Peng, Y., Sun, H.: VLUCI: variational learning of unobserved confounders for counterfactual inference (2023)

Causal Inference in the Multiverse
of Hazard

En-Yu Lai⬤ and Yen-Tsung Huang$^{(\boxtimes)}$⬤

Institute of Statistical Science, No. 128, Academia Road, Section 2, Nankang, Taipei
115201, Taiwan
ythuang@stat.sinica.edu.tw

Abstract. Hazard is a crucial estimand in both applied and methodological contexts. However, its causal interpretation is challenging due to
inherent selection biases and the ambiguity in defining populations for
comparison across different treatment groups. To address these issues,
we introduce a novel definition of counterfactual hazard based on the
framework of possible worlds. Rather than conditioning on prior survival
status as a conditional probability, our definition involves intervening in
the prior status, treating it as a marginal probability. Using single-world
intervention graphs, we show that the proposed counterfactual hazard
represents a controlled direct effect. Conceptually, intervening in survival
status at each time point creates a new possible world. The proposed hazards at these time points represent risks in these hypothetical scenarios,
forming a "multiverse."

Keywords: counterfactual process · hazard · survival analysis ·
possible-world semantics

1 Introduction

Hazard plays a vital role in both applied and methodological research. Tradi-
tionally defined as a conditional probability, it represents the likelihood of expe-
riencing an outcome at time t given its absence up to time t^-. This conditional
perspective holds practical appeal, often surpassing the marginal probability.
For instance, amid the COVID-19 pandemic, individuals may be more concerned
about their probability of contracting the virus, given their prior health status,
than the overall marginal probability since December 2019. Methodologically,
hazard stands as a fundamental estimand in classic life table analyses [1] and
serves as a crucial intermediary in the Nelson-Aalen [2] and Kaplan-Meier [3]
estimators. In regression modeling, hazard-based association measures can also
be conveniently obtained by the Cox proportional hazards model [4] and Aalen's
additive hazards model [5].

Despite the practical and methodological value of hazard, concerns regarding its causal interpretation have been widely acknowledged [6–9]. Firstly, hazard inherently carries a selection bias. Even in randomized studies, the balance

between treatment and placebo groups may be disrupted over time due to differential depletion of susceptibles, particularly if the treatment has a causal effect [6]. Secondly, hazard-based effect estimates compares different populations, which leads to difficulty in attributing causality to a well-defined population, even in a counterfactual framework [8,9]. Lastly, hazard is a non-collapsible quantity, meaning that the marginal hazard may not be recovered by a weighted average of conditional hazards [7].

This paper aims to establish a conceptual framework for hazard by introducing a pseudo-population concept across the multiverse [10]. We propose a counterfactual hazard as a marginal probability (Sect. 2.1). We delineate its underlying counterfactual population in the multiverse under a possible world framework and elucidate an interpretation of the proposed hazard as risks in the pseudo-populations of the multiverse (Sect. 3). The proposed counterfactual hazard can further be summed up across the multiverse as a cumulative hazard or averaged as an average hazard. We also show that the risk in the actual world is bounded by the average and cumulative hazards (Sect. 4).

2   Counterfactual Hazards: cCT and iCP-Hazards

2.1   Definitions and Assumptions

We let Z_i, T_i, and C_i, respectively, denote the treatment, survival time, and censoring time for subject i where $i = 1, ..., m$ and m is the sample size and assume that T_i and C_i share the same origin and are independent conditional on Z_i and X_i, i.e., noninformative censoring, where X_i are measured covariates. We also define the following processes: $\tilde{N}_i(t) = I(T_i \leq t)$ denotes the underlying event process where $I(\cdot)$ is an indicator function; $N_i(t) = I(T_i \leq t, T_i \leq C_i)$ denotes the observed event process; $Y_i(t) = I(T_i \geq t, C_i \geq t)$ denotes the at-risk process. We further introduce notations based on counterfactuals. $T_i(z)$ denotes counterfactual survival time for subject i whose Z_i had been set to z, and $\tilde{N}_i(t; z, n(t^-))$ denotes a counterfactual event process [11] at time t had Z_i and $\tilde{N}_i(t^-)$ been set to z and $n(t^-)$, respectively, where $n(t^-) \in \{0,1\}$. The intervention $\tilde{N}_i(t^-) = n(t^-)$ creates a new possible world at each t that will be illustrated in Sect. 3.

We introduce two counterfactual hazards, with one based on the counterfactual survival time $T(z)$ (cCT) and the other using the counterfactual event process (iCP):

$$d\Lambda_{cCT}(t \mid z) := P\{T(z) \in [t, t + dt) \mid T(z) \geq t\} \tag{1}$$

$$d\Lambda_{iCP}(t \mid z) := P\{\tilde{N}(t; z, n(t^-)) = 0) = 1\}. \tag{2}$$

The cCT (conditioning counterfactual survival time) defines the hazard by conditioning on the event of $T(z) \geq t$ whereas the iCP (interventional counterfactual process) does so by an intervention setting $\tilde{N}(t^-) = n(t^-) = 0$. Note that $n(t^-)$

represents an intervention $\in \{0, 1\}$ depending on time. We further denote the two hazards conditional on covariates $X = x$:

$$d\Lambda_{cCT}(t \mid z, x) := P\{T(z) \in [t, t + dt) \mid T(z) \geq t, X = x\}$$
$$d\Lambda_{iCP}(t \mid z, x) := P\{\tilde{N}(t; z, n(t^-)) = 0) = 1 \mid X = x\}.$$

By the definition of $d\Lambda_{iCP}(t \mid z)$, it follows that

$$d\Lambda_{iCP}(t \mid z) = \sum_x d\Lambda_{iCP}(t \mid z, x) P(X = x)$$
$$= \sum_x P(d\tilde{N}(t) = 1 \mid \tilde{N}(t^-) = 0, Z = z, X = x) P(X = x),$$

which, however, is different from that for $dA_{cCT}(t \mid z)$ (derived under $T(z) \perp\!\!\!\perp Z \mid X$):

$$dA_{cCT}(t \mid z) = \frac{\sum_x P(T \in [t, t + dt) \mid Z = z, X = x) P(X = x)}{\sum_x P(T \geq t \mid Z = z, X = x) P(X = x)}$$
$$\neq \sum_x dA_{cCT}(t \mid z, x) P(X = x),$$

due to the non-collapsibility of a conditional probability. Under noninformative censoring, we can construct the following estimators

$$d\hat{\Lambda}_{cCT}(t \mid z) = \frac{\sum_x \sum_{i=1}^m dN_i(t) I(X_i = x, Z_i = z) \sum_{i=1}^m I(X_i = x)}{\sum_x \sum_{i=1}^m Y_i(t) I(X_i = x, Z_i = z) \sum_{i=1}^m I(X_i = x)} \tag{3}$$

$$d\hat{\Lambda}_{iCP}(t \mid z) = \sum_x \frac{\sum_{i=1}^m dN_i(t) I(X_i = x, Z_i = z)}{\sum_{i=1}^m Y_i(t) I(X_i = x, Z_i = z)} \frac{1}{m} \sum_{i=1}^m I(X_i = x). \tag{4}$$

2.2   Graphical Illustration

Given the complexity of depicting counterfactual survival time in a time-dependent manner on causal diagrams, our focus here is on the iCP hazard, where the process can be discretized into a series of Bernoulli variables, such as $\tilde{N}(t_1)$, $\tilde{N}(t_2)$, and so forth. We illustrate the iCP hazard using single-world intervention graphs (SWIGs) [12]. Comparing a counterfactual hazard between different values of Z may be viewed as a direct causal effect of Z on the outcome not through the earlier outcomes. This is achieved by the iCP via setting $\tilde{N}(t^-) = n(t^-) = 0$ (depicted by the blue arrow in Fig. 1(a)). Adjustments for the exposure-outcome confounder are necessary to ensure $\tilde{N}(t; z, n(t^-)) \perp\!\!\!\perp Z \mid X$ (Fig. 1(b)), and adjustments for the outcome-outcome confounder (Fig. 1(c)) are needed to ensure $\tilde{N}(t; z, n(t^-)) \perp\!\!\!\perp \tilde{N}(t^-) \mid X$. Similar to identifying controlled direct effects in mediation analyses [13,14], the iCP hazard is identifiable in the presence of treatment-induced outcome-outcome confounder that is construed as a type of outcome-outcome confounders.

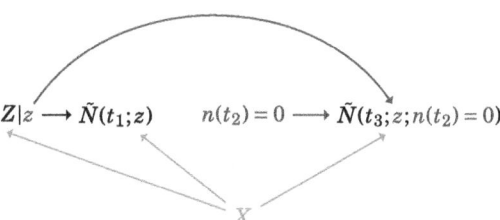

(a) iCP as a controlled direct effect

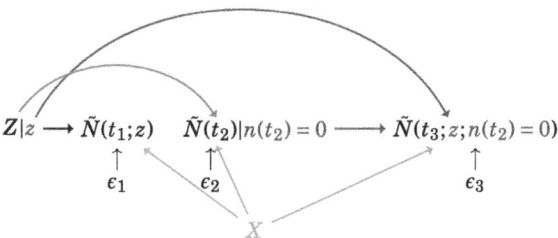

(b) exposure-outcome confounder needs to be adjusted

(c) outcome-outcome confounder needs to be adjusted

Fig. 1. Graphical illustrations of the iCP hazard.

3　Multiverse of Hazards

In this paper, we leverage elements from possible-world semantics [15], a classical method used to determine the truth value of counterfactual conditionals. We adapt these concepts to develop a customized semantics for counterfactual hazards, which we refer to as the *multiverse semantic*. In our framework, the *actual world* represents the reality that we observe events and collect data from. A *possible world* is a virtual counterpart to the actual world, where we can introduce artificial settings to calculate statistics with causal interpretation. Within this framework, an *agent* refers to a statistician who observes events and computes statistics through these worlds.

3.1　Direct Interpretation

As illustrated in Fig. 2 (first panel), participants surviving in the control group tend to be inherently healthier than those in the treatment group over time, assuming treatment benefits. The core concept behind the iCP-hazard is to compute hazards from possible worlds where control and treatment groups remain comparable. This world is achieved by intervening with $n(t^-) = 0$, effectively reviving all deceased individuals at t^- in the actual world. In this possible world, those who die at t in the actual world will also perish, while those revived at t will experience their potential outcomes that they would not have had in the actual world. Using this semantics, the agent calculates hazard at each time point t by simulating a scenario where all deceased individuals are revived by t^- in the actual world. This approach is depicted at the individual level in Fig. 2 (second panel).

3.2 Multiverse Interpretation

As depicted in Fig. 3 (upper panel), the actual risk evaluates the same population and total deaths within a specific time frame. However, the cumulative iCP-hazard considers different pseudo-populations across various time slices, where the sample size fluctuates throughout the follow-up due to the die-and-revive intervention. To circumvent the need for reviving interventions, we propose an alternative interpretation for the iCP-hazard: the multiverse interpretation. This approach provides a clearer conceptualization of the pseudo-population underlying the cumulative iCP-hazard.

Instead of individually reviving the deceased at each event time (the direct interpretation), we establish a series of counterfactual scenarios from the outset, identical to the actual world at t_0. In each of these possible worlds, individuals can only perish at a predetermined time point, as depicted in Fig. 2 (third panel). This ensemble of possible worlds is termed the *multiverse of hazard*. Under this framework, the j-th event time, denoted as t_j, in the actual world corresponds to the j-th possible world, created by setting $n(t_j^-) = 0$, allowing the agent to compute the hazard in each scenario. Similar to the direct interpretation, individuals who perish at t_j in the actual world will also perish in the j-th possible world. Those who die before t_j in the actual world receive their potential outcomes in the j-th possible world, outcomes not observed had they died earlier in the actual world. Since no deaths occur before t_j in the j-th possible world, the comparability of control and treatment groups at t^- mirrors that at baseline. Consequently, the hazard from each possible world inherits a valid causal interpretation, provided the two groups are randomized at baseline. Moreover, since no deaths occur after t_j, the hazard represents the risk in each possible world.

In summary, Fig. 3 (boxed panel) illustrates three distinct routes for the agent. The black path signifies the conventional hazard assessment, where the agent remains in the actual world throughout the follow-up, observing cross-time survival statuses and calculating both classical hazards and actual risk. The gray path represents the direct interpretation of the iCP-hazard. Here, the agent traverses between actual and possible worlds, observing cross-world survival statuses and computing the counterfactual hazard. On the other hand, the green, red, and yellow paths depict the multiverse interpretation of the iCP hazard. In this scenario, the agent journeys within each possible world, tracking survival statuses over time and computing the counterfactual hazard as a risk. Conceptually, this statistician agent takes on a super-hero task similar to that of Dr. Strange in Marvel.

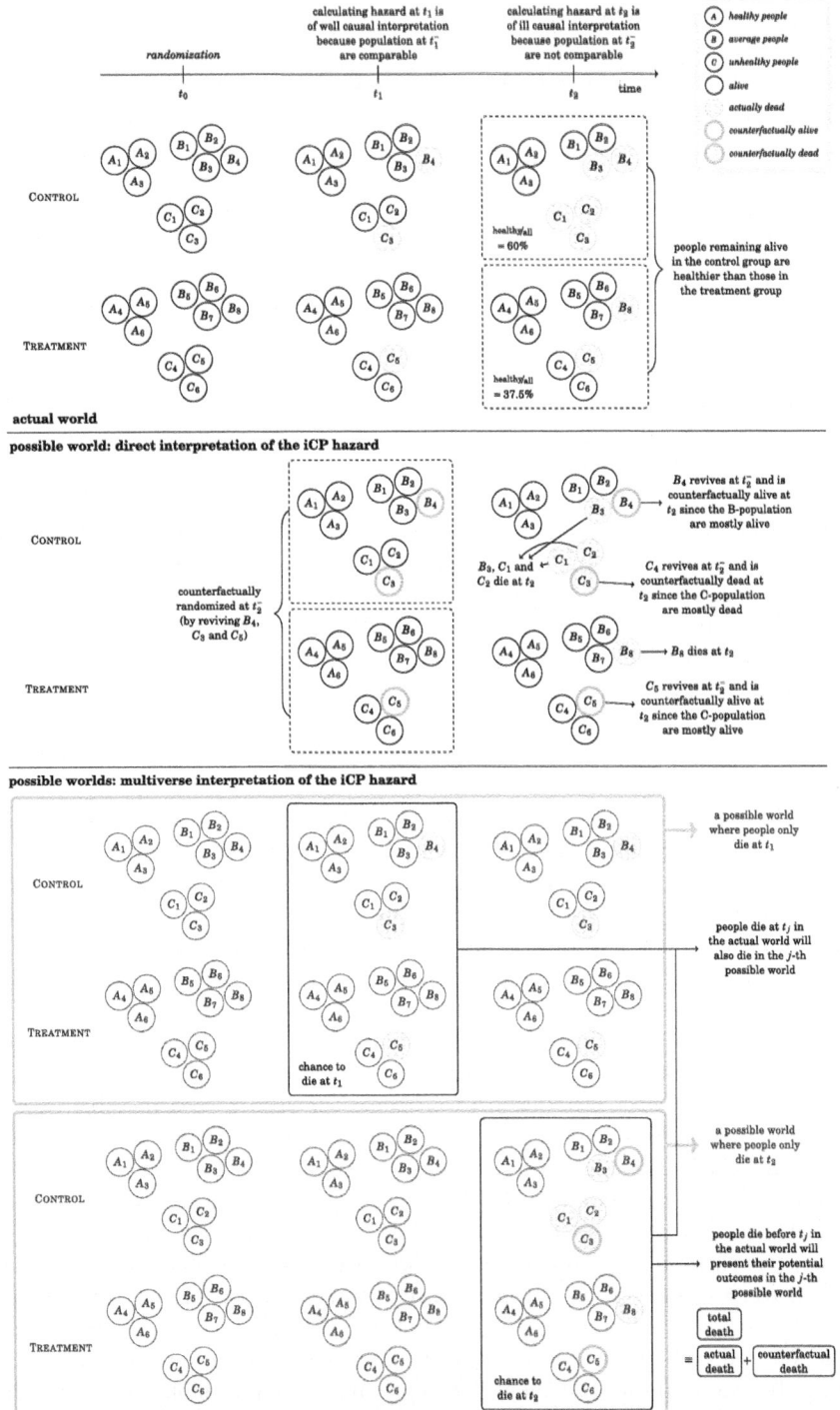

Fig. 2. Multiverse interpretation of the iCP hazard.

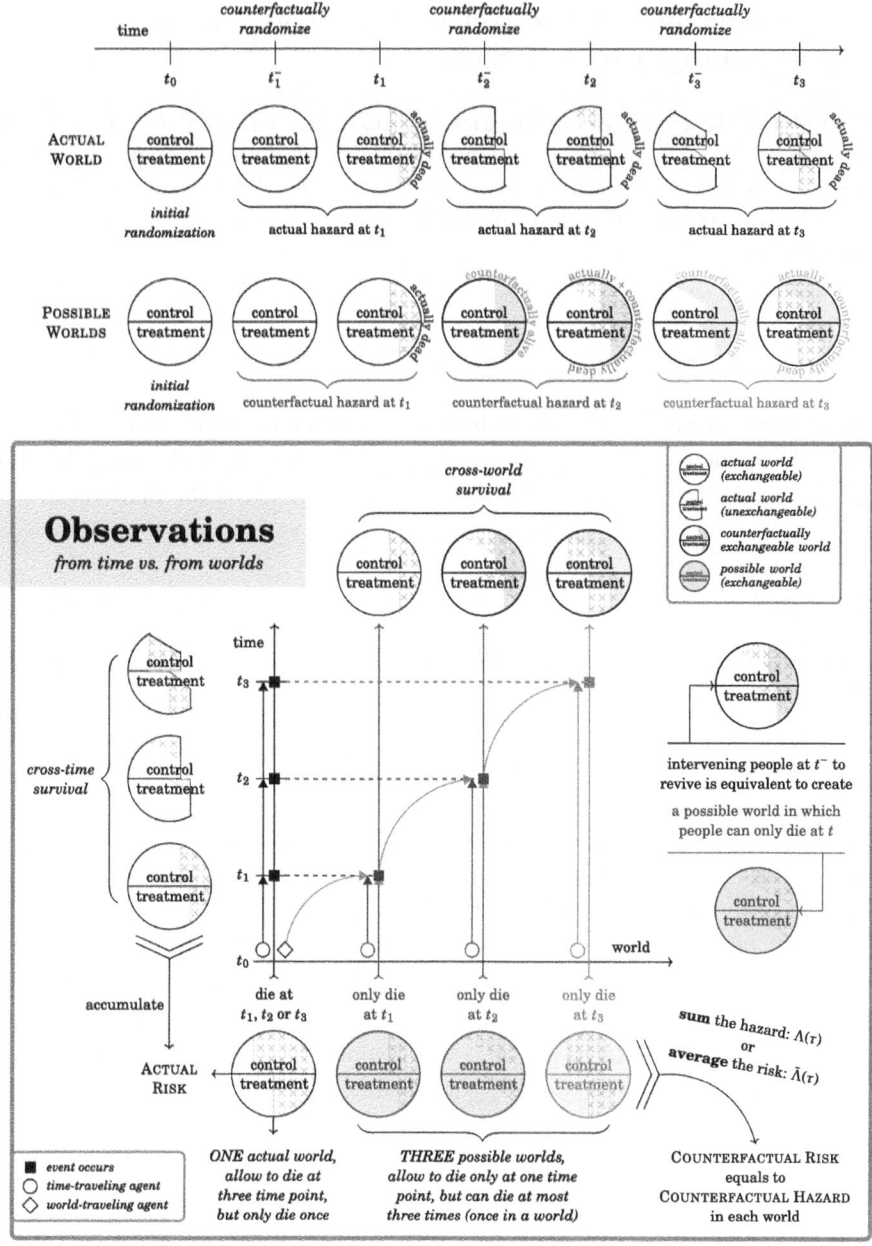

Fig. 3. Multiverse interpretation of the iCP cumulative hazard.

4   Relationships of the Actual Risk and (cumulative and Average) iCP-Hazard

We omit the treatment Z for succinct notations and consider J discrete event time points: $t_1 < \ldots < t_J < \tau$, where τ marks the study's end. The *cumulative counterfactual hazard*, denoted $\Lambda_{iCP}(\tau)$, sums the ratio of total deaths d_j (including actual and potential deaths) in the j-th possible world over the sample size m:

$$\Lambda_{iCP}(\tau) := \sum_{j=1}^{J} \frac{d_j}{m},$$

Its estimation $\hat{\Lambda}_{iCP}(\tau)$ aggregates the counterfactual hazard estimates $\hat{\Lambda}_{iCP}(t_j)$ across the event times t_j: $\hat{\Lambda}_{iCP}(\tau) = \sum_{j=1}^{J} d\hat{\Lambda}_{iCP}(t_j)$. Similarly, we define the *average counterfactual hazard*, $\bar{\Lambda}_{iCP}(\tau)$, as the mean across the multiverse:

$$\bar{\Lambda}_{iCP}(\tau) := \frac{1}{J} \sum_{j=1}^{J} \frac{d_j}{m}.$$

Contrastingly, the risk in the actual world, denoted $F(\tau)$, sums the actual deaths $d\tilde{N}(t_j)$ over the sample size:

$$F(\tau) := \frac{\sum_{j=1}^{J} d\tilde{N}(t_j)}{m},$$

Under the assumption that deaths at time t_j in the actual world correspond to deaths in the j-th possible world, $d\tilde{N}(t_j) \leq d_j$, with equality if all deaths occur simultaneously or at the first event time (i.e., $j = 1$).

In any possible world within the multiverse, the number of deaths does not exceed the total deaths in the actual world, $d_j \leq \sum_{j=1}^{J} d\tilde{N}(t_j)$. It follows that $\sum_{j=1}^{J} d_j \leq \sum_{j=1}^{J} \sum_{j=1}^{J} d\tilde{N}(t_j) = J \sum_{j=1}^{J} d\tilde{N}(t_j)$. Conversely, as previously explained, $d\tilde{N}(t_j) \leq d_j$. Combining these inequalities, we derive $J^{-1} \sum_{j=1}^{J} d_j \leq \sum_{j=1}^{J} d\tilde{N}(t_j) \leq \sum_{j=1}^{J} d_j$, leading to:

$$\bar{\Lambda}_{iCP}(\tau) \leq F(\tau) \leq \Lambda_{iCP}(\tau).$$

These inequalities indicate that the risk in the actual world lies between the average iCP hazard and the cumulative iCP hazard, serving as lower and upper bounds, respectively.

5   Discussion

This paper has the following contributions. Firstly, we introduce the iCP-hazard, a counterfactual hazard expressed as a marginal probability via a counterfactual process. Unlike traditional approaches that condition on prior survival status, the iCP-hazard intervenes in this status instead. While the identification formula of the iCP-hazard, conditional on covariates, aligns with that derived from the cCT-hazard defined by counterfactual survival time, the two differ in their handling of covariates due to the non-collapsibility of the cCP-hazard. Additionally, compared to counterfactual survival time, the process-based hazard is more straightforward to illustrate in a causal diagram. Secondly, we establish an underlying pseudo-population for the counterfactual iCP-hazard. Specifically, we conceptualize a multiverse expanded by the number of event times, where the risk in each possible world corresponds to the hazard at each time point. Lastly, by linking hazards with risks in the multiverse, we enable causal inference that draws from developments in counterfactual risk, potentially addressing issues associated with hazard analysis.

We have demonstrated that the proposed iCP hazard functions as a controlled direct effect, with the prior outcome status acting as the mediator. Identification of this hazard relies on the assumption of no unmeasured confounding: $\tilde{N}(t; z, n(t^-)) \perp\!\!\!\perp (Z, \tilde{N}(t^-)) \mid X$. As depicted in Fig. 1, adjustments must be made for treatment-outcome and outcome-outcome confounders. The latter may vary over time, as the confounders for $\tilde{N}(t_1)$ and $\tilde{N}(t_2)$ may differ from those for $\tilde{N}(t_3)$ and $\tilde{N}(t_4)$. In practice, identifying and accounting for all time-dependent confounders is challenging. Even with available data, proper adjustment for time-varying confounding is essential to ensure a valid causal interpretation [16]. When the assumption of no unmeasured confounding is violated, alternative strategies can be employed to partially identify the iCP hazard. For example, if a monotonic relationship exists between the unmeasured confounder and the treatment, mediator, and outcome, bounds have been established for the controlled direct effect [17]. Alternatively, in scenarios where information about the unmeasured confounder is unavailable, two instruments—one for the treatment and another for the mediator—may be utilized to identify the controlled direct effect among compliers [18].

Acknowledgments. We thank the Ministry of Science and Technology, Taiwan (108-2118-M-001-013-MY5) and Academia Sinica (AS-CDA-108-M03) for providing funding for this study.

Disclosure of Interests. The authors have no competing interests in declaring the content relevant to this article.

Full version. A comprehensive version with data application is available at https://arxiv.org/abs/2405.04446.

References

1. Keyfitz, N., Frauenthal, J.: An improved life table method. Biometrics, 889–899 (1975)
2. Aalen, O.O.: Nonparametric estimation for a family of counting processes. Ann. Stat. **6**, 701–726 (1978)

3. Kaplan, E.L., Meier, P.: Nonparametric estimation from incomplete observation. J. Am. Stat. Assoc. **53**, 457–481 (1958)
4. Cox, D.R.: Regression models and life-tables. J. Roy. Stat. Soc. B **34**, 187–220 (1972)
5. Aalen, O.O.: A linear regression model for the analysis of life times. Stat. Med. **8**, 907–925 (1989)
6. Hernán, M.A.: The hazards of hazard ratios. Epidemiology **21**, 13–15 (2010)
7. Aalen, O.O., Cook, R.J., Roysland, K.: Does cox analysis of a randomized survival study yield a causal treatment effect? Lifetime Data Anal. **21**, 579–593 (2015)
8. Martinussen, T., Vansteelandt, S., Andersen, P.K.: Subtleties in the interpretation of hazard contrasts. Lifetime Data Anal. **26**, 833–855 (2020)
9. Didelez, V., Stensrud, M.J.: On the logic of collapsibility for causal effect measures. Biom. J. **64**, 235–242 (2022)
10. Bousso, R., Susskind, L.: Multiverse interpretation of quantum mechanics. Phys. Rev. D **85**, 045007 (2012)
11. Huang, Y.T.: Causal mediation of semicompeting risks. Biometrics **77**, 1143–1154 (2021)
12. Richardson, T.S., Robins, J.M.: Single world intervention graphs (swigs): a unification of theh counterfactual and graphical approaches to causality (2013)
13. VanderWeele, T.J., Vansteelandt, S.: Conceptual issues concerning mediation, interventions and composition. Stat. Inference **2**, 457–468 (2009)
14. Imai, K., Keele, L., Yamamooto, T.: Identification, inference and sensitivity analysis for causal mediation effects. Stat. Sci. **25**, 51–71 (2009)
15. Lewis, D.K.: Counterfactuals. Mass, Blackwell, Malden (1973)
16. Robins, J.M., Hernán, M.A., Brumback, B.: Marginal structural models and causal inference in epidemiology. Epidemiology **11**, 550–560 (2000)
17. VanderWeele, T.J.: Controlled direct and mediated effects: definition, identification and bounds. Scandinavian J. Stat. **38**(3), 551–563 (2011)
18. Frölich, M., Huber, M.: Direct and indirect treatment effects-causal chains and mediation analysis with instrumental variables. J. R. Stat. Soc. Ser. B Stat Methodol. **79**(5), 1645–1666 (2017)

A Continuous Structural Intervention Distance to Compare Causal Graphs

Mihir Dhanakshirur[1], Felix Laumann[2]([✉]), Junhyung Park[3],
and Mauricio Barahona[2]

[1] Department of Mathematics, Indian Institute of Science, Bengaluru,
Bangalore, India
`mihird@iisc.ac.in`
[2] Department of Mathematics, Imperial College London, London, UK
`f.laumann18@imperial.ac.uk`
[3] MPI for Intelligent Systems, Tübingen, Germany

Abstract. Causal inference under interventions requires accurate assessment of differences between true and learned causal graphs. We introduce a new continuous metric that extends beyond graph-based measures like Structural Hamming Distance and Structural Intervention Distance by incorporating underlying data alongside graph structures. Our approach embeds intervention distributions for each node pair as conditional mean embeddings in reproducing kernel Hilbert spaces, then quantifies their disparity using maximum (conditional) mean discrepancy. We present theoretical findings supported by synthetic data experiments.

Keywords: Directed acyclic graphs · graph distance · conditional mean embeddings · kernel methods

1 Introduction

In many causal learning settings, it is assumed that data are generated according to a Structural Causal Model (SCM) that encapsulates causal relationships between variables. Under the faithfulness and causal Markov assumptions, the distribution entailed by an SCM is directly related to an underlying Directed Acyclic Graph (DAG), which will be called the *true* DAG. The task in causal learning is to infer a *learnt* DAG, given access to data generated by the underlying SCM. Here, we are concerned with the problem of quantifying the performance of causal structure learning algorithms by measuring the distance between the two DAGs (learnt and true) taking into account both graph structure and the presence of any edge weights.

M. Dhanakshirur and F. Laumann—Equal contribution.

Supplementary Information The online version contains supplementary material available at https://doi.org/10.1007/978-981-97-7812-6_3.

X.-H. Zhou and J. Jia (Eds.): PCIC 2024, CCIS 2200, pp. 25–40, 2025.
https://doi.org/10.1007/978-981-97-7812-6_3

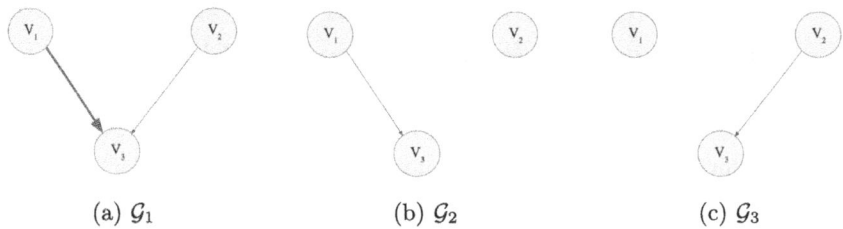

Fig. 1. The true DAG $\mathcal{G}_1$ and two learnt DAGs, $\mathcal{G}_2$ and $\mathcal{G}_3$, where the increased width of the edge from V_1 to V_2 in $\mathcal{G}_1$ signalises its greater weight.

Although a variety of metrics exist to compare causal graphs [e.g., 1, 2, 5, 10, 12, 15]. the most prominent ones, the Structural Hamming Distance (SHD) and the Structural Intervention Distance (SID), are dominated by graph properties alone. The SHD is the square of the Frobenius norm of the difference between the binary adjacency matrices of the true and learnt DAGs; hence it directly counts the number of edges to be added and removed to the learnt DAG so that it is equal to the true DAG. The SID also provides a count; in this case of the number of pairwise interventional distributions on which the true DAG and the learnt DAG differ.

In this work, we propose a distance, denoted the *continuous Structural Intervention Distance* (contSID), which is aimed at quantifying distances between causal graphs by taking into account both graph structure and data-dependent edge weights. Given that the goal of learning a DAG in causal learning is to later use it to estimate interventional effects, it is important to go beyond mere counts, as in SHD or SID, and quantify the difference in interventional distributions from the two DAGs, which are affected by edge weights.

To illustrate this issue, consider data generated by the following linear model:

$$V_1 \sim \mathcal{N}(0,1), \quad V_2 \sim \mathcal{N}(0,1), \quad V_3 \sim \mathcal{N}(10\,V_1 + V_2, 1), \qquad (1)$$

which is associated with the true DAG $\mathcal{G}_1$ (Fig. 1a), where the edge connecting V_1 and V_3 has a mean weight of 10. Suppose that $\mathcal{G}_2$ and $\mathcal{G}_3$ (Figs. 1b–1c) are learnt DAGs from two different causal discovery algorithms. Intuitively, one expects that missing the edge $V_1 \rightarrow V_3$ should be penalised more than missing the edge $V_2 \rightarrow V_3$, since an intervention on V_1 would lead to a larger difference in the distribution of V_3 than the same intervention on V_2. Yet Table 1 shows that the SHD and SID fail to distinguish the quality of the two learnt DAGs relative to the true DAG. In contrast, the contSID defined here is able to capture the differences in the interventional distributions by computing the norm difference of their interventional mean embeddings in a reproducing kernel Hilbert space (RKHS). We use the observational distribution (via the valid adjustment set) to estimate their mean embeddings.

Table 1. SHD, SID and contSID calculated on $d(\mathcal{G}_1, \mathcal{G}_2)$ and $d(\mathcal{G}_1, \mathcal{G}_3)$.

Metric	$d(\mathcal{G}_1, \mathcal{G}_2)$	$d(\mathcal{G}_1, \mathcal{G}_3)$
SHD	1	1
SID	1	1
contSID	0.23	0.39

2 Background

We consider a finite set of random variables $X_1, \ldots, X_D$ with index set $\mathbf{V} = \{1, \ldots, D\}$. A graph $\mathcal{G} = (\mathbf{V}, \mathcal{E})$ consists of nodes (or vertices) $\mathbf{V}$ and edges $\mathcal{E} \subseteq \mathbf{V} \times \mathbf{V}$. We will identify a node $j \in \mathbf{V}$ with its corresponding random variable X_j; hence we will use variables, nodes and vertices interchangeably depending on the context. We denote random variables by upper-case and their corresponding samples by lower-case letters, e.g., the sample of the random variable X_i is denoted by x_i. The parent set of node X_i is denoted by $\mathbf{PA}_i := \{X_j | (i, j) \in \mathcal{E}, j \in \mathbf{V}\}$ and its sample is $\mathbf{pa}_i$.

We assume that the observational data $\mathcal{D} = \{x_1^{(n)}, \ldots, x_D^{(n)}\}_{n=1}^N$ are a set of N independent samples from a D-dimensional joint distribution P with density $p(\cdot)$ with respect to the Lebesgue or counting measure. Additionally, we require that the distribution is Markov with respect to a directed acyclic graph (DAG) $\mathcal{G}$ (see Appendix A for definitions) according to the following definition.

Definition 1 (Causal Markov assumption [11]). *The distribution P is said to satisfy the Markov factorization property with respect to the DAG $\mathcal{G}$ if*

$$p(x_1, \ldots, x_D) = \prod_{j=1}^{D} p(x_j | \mathbf{pa}_j) \tag{2}$$

2.1 Interventional Distribution and *do*-Calculus

Given the causal DAG, $\mathcal{G}$, and random variables X_i and X_j, our aim is to estimate the interventional distribution of X_j, $P_{X_j | do(X_i) = \hat{x}_i}$, where $do(X_i) = \hat{x}_i$ represents an intervention on X_i that sets its value to $\hat{x}_i$.

Definition 2 (Interventional distribution). *The density of the interventional distribution given $do(X_i) = \hat{x}_i$ is defined as*

$$p(x_1, \ldots, x_D | do(X_i) = \hat{x}_i) = \prod_{j \neq i}^{D} p(x_j | \mathbf{pa_j}) \, \delta(x_i = \hat{x}_i) , \tag{3}$$

where $\delta(x_i = \hat{x}_i)$ is the distribution X_i after we intervened on it.

Typically, the interventional distribution is not accessible by experiment, since we are usually only given observational data. The *do*-calculus [9] enables us to estimate interventional distributions from observational distributions from a given DAG through valid adjustment sets [10].

Definition 3 (Valid adjustment set). *Let $X_j \notin \mathbf{PA}_i$ (see Remark 1). A set $\mathbf{Z} \subseteq \mathbf{V} \setminus \{i, j\}$ is a valid adjustment set for the ordered pair (X_i, X_j) if*

$$p(x_j|do(X_i) = \hat{x}_i) = \int_{\mathcal{Z}} p(x_j|\hat{x}_i, \mathbf{z}) \, p(\mathbf{z}) d\mathbf{z}. \tag{4}$$

where $\mathcal{Z}$ is the non-empty set in which $\mathbf{Z}$ takes values.

Remark 1. If $X_j \in \mathbf{PA}_i$, the marginal distribution of X_j obtained from the observational distribution (2) and the interventional distribution (3) are identical. So, $P_{X_j|do(X_i)=\hat{x}_i}$ is equal to the observational distribution of X_j for every $\hat{x}_i$; hence interventions have no effect.

The following theorem provides a full characterisation of valid adjustment sets.

Theorem 1 (Characterisation of valid adjustment sets [10,14]). *Given the causal DAG, $\mathcal{G}$, and the ordered pair of variables (X_i, X_j), a subset $\mathbf{Z} \subseteq \mathbf{V} \setminus \{i, j\}$ is a valid adjustment set for $P_{X_j|do(X_i)}$ if it satisfies the following two properties:*

(i) $\mathbf{Z}$ *blocks all undirected paths in $\mathcal{G}$ between X_i and X_j*
(ii) no $Z \in \mathbf{Z}$ is a descendant of any X_k that lies on a directed path from X_i to X_j (except for any descendants of X_i that are not on a directed path from X_i to X_j).

Remark 2. For an ordered pair of nodes (X_i, X_j) there exist many valid adjustment sets. For instance, $\mathbf{PA}_i$, the parent set of X_i, is an example of a valid adjustment set but many more may exist.

2.2   Kernel-Based Representations for Distributions

Henceforth, let $(\Omega, \mathcal{F}, \mathcal{P})$ be the underlying probability space, let $(\mathcal{X}, \mathfrak{X})$ and $(\mathcal{Z}, \mathfrak{Z})$ be separable measurable spaces, and let $X : \Omega \to \mathcal{X}$ and $Z : \Omega \to \mathcal{Z}$ be random variables with distributions P_X and P_Z. We will also consider random variables $X' : \Omega \to \mathcal{X}$ and $Z' : \Omega \to \mathcal{Z}$. Let $\mathcal{H}_{\mathcal{X}}$ be a vector space of functions $f : \mathcal{X} \to \mathbb{R}$ endowed with a Hilbert space structure via an inner product $\langle \cdot, \cdot \rangle_{\mathcal{H}_{\mathcal{X}}}$.

A mean embedding is a mapping of a probability distribution onto an RKHS by a kernel k. This mapping is one-to-one if the kernel is characteristic [4], as described by the following definitions.

Definition 4 (Reproducing kernel). *A symmetric function $k_{\mathcal{X}} : \mathcal{X} \times \mathcal{X} \to \mathbb{R}$ is a reproducing kernel of $\mathcal{H}_{\mathcal{X}}$ if and only if:*

(i) $\forall x \in \mathcal{X}$, $k_{\mathcal{X}}(x, \cdot) \in \mathcal{H}_{\mathcal{X}}$, and

(ii) $\forall x \in \mathcal{X}$ and $\forall f \in \mathcal{H}_{\mathcal{X}}$, $f(x) = \langle f, k_{\mathcal{X}}(x, \cdot) \rangle_{\mathcal{H}_{\mathcal{X}}}$.

The Hilbert space associated with a reproducing kernel is known as a Reproducing Kernel Hilbert Space (RKHS).

Definition 5 (Kernel mean embedding). *Given a distribution P_X on $\mathcal{X}$ and assuming $\mathbb{E}_X[k_{\mathcal{X}}(X, X)] < \infty$, the kernel mean embedding of P_X is defined as*

$$\mu_{P_X}(\cdot) := \mathbb{E}_X[k_{\mathcal{X}}(X, \cdot)] \tag{5}$$

Definition 6 (Characteristic kernel). *A positive definite kernel $k_{\mathcal{X}}$ is characteristic to a set $\mathcal{P}$ of probability measures on $\mathcal{H}_{\mathcal{X}}$ if the map $\mathcal{P} \to \mathcal{H}_{\mathcal{X}}$ given by $P_X \mapsto \mu_{P_X}$ is injective.*

Remark 3. Popular choices like the Gaussian and Laplacian kernels are characteristic [18].

The RKHS associated with a characteristic kernel enables us to distinguish between different distributions by defining the maximum mean discrepancy, a distance between the kernel mean embeddings of the two distributions.

Definition 7 (Maximum mean discrepancy). *The maximum mean discrepancy (MMD) between two distributions $P_X, P_{X'} \in \mathcal{P}$ is defined as*

$$MMD_{P_X, P_{X'}} = ||\mu_{P_X} - \mu_{P_{X'}}||_{\mathcal{H}_{\mathcal{X}}} \tag{6}$$

The MMD is a measure of discrepancy between distributions that is widely-used in the machine learning literature due to its theoretical properties and ease of empirical estimation. Although the MMD forms the backbone of our approach in this paper, we note, however, that there are several other such measures, and leave it as a future research direction to investigate how those can be utilised for the problems we tackle in this paper.

Our problem setting for interventional distributions is based on the evaluation of conditional distributions; hence we recap some general results for conditional mean embeddings [8]. Note that, rather than following the standard operator-based definitions [16], we adopt here the measure-theoretic approach to kernel conditional mean embeddings as in [8]. This approach has some advantages: it does not rely on stringent assumptions for the population version of the embedding to exist; and it is endowed with a natural regression interpretation for empirical estimates.

Definition 8 (Conditional mean embedding [8]). *Assuming that $\mathbb{E}_X[k_{\mathcal{X}}(X, X)] < \infty$, the conditional mean embedding (CME) of $P_{X|Z}$ is defined as:*

$$\mu_{P_{X|Z}} := \mathbb{E}_{X|Z}[k_{\mathcal{X}}(X, \cdot)|Z]. \tag{7}$$

Hence $\mu_{P_{X|Z}}$ is a Z-measurable random variable taking values in $\mathcal{H}_{\mathcal{X}}$.

The following theorem provides the basis for estimating the CME of the conditional distribution $P_{X|Z}$.

Theorem 2 (Deterministic function of conditional mean embedding [8]). *Let $\mathcal{B}(\mathcal{H}_\mathcal{X})$ be the Borel σ-algebra of $\mathcal{H}_\mathcal{X}$, then the CME can be written as:*

$$\mu_{P_{X|Z}} = F_{P_{X|Z}} \circ Z,$$

where $F_{P_{X|Z}} : \mathcal{Z} \to \mathcal{H}_\mathcal{X}$ is a deterministic function measurable with respect to $\mathfrak{Z}$ and $\mathcal{B}(\mathcal{H}_\mathcal{X})$ and $\circ$ denotes composition of functions.

We can now define the analogue to the MMD for conditional distributions $P_{X|Z}$ and $P_{X'|Z'}$, which is denoted the *maximum conditional mean discrepancy*.

Definition 9 (Maximum Conditional Mean Discrepancy [8]). *In addition to the random variables X and Z, consider also the random variables $X' : \Omega \to \mathcal{X}$ and $Z' : \Omega \to \mathcal{Z}$. Assuming that $E_X[k_\mathcal{X}(X,X)] < \infty$ and $E_{X'}[k_\mathcal{X}(X',X')] < \infty$. Then the maximum conditional mean discrepancy (MCMD) between $P_{X|Z}$ and $P_{X'|Z'}$ is the function from $\mathcal{Z} \to \mathbb{R}$ defined by*

$$MCMD_{P_{X|Z},P_{X'|Z'}}(z) = ||F_{P_{X|Z}}(z) - F_{P_{X'|Z'}}(z)||_{\mathcal{H}_\mathcal{X}} \tag{8}$$

Remark 4. From Eq. (7) and Theorem 2, it follows that the kernel mean embedding of the distribution $P_{X|Z=z}$ is given by

$$\mu_{P_{X|Z=z}} = \mathbb{E}_X[k_\mathcal{X}(X,\cdot)|Z=z] = F_{P_{X|Z}}(z). \tag{9}$$

Therefore it follows that the MCMD (8) at z is equal to the MMD between the distributions $P_{X|Z=z}$ and $P_{X'|Z'=z}$:

$$\text{MCMD}_{P_{X|Z},P_{X'|Z'}}(z) = ||\mu_{P_{X|Z=z}} - \mu_{P_{X'|Z'=z}}||_{\mathcal{H}_\mathcal{X}} = \text{MMD}_{P_{X|Z=z},P_{X'|Z'=z}}. \tag{10}$$

We use this fact below to construct a plug-in estimate of the MCMD.

By Theorem 2, the task of estimating $\mu_{P_{X|Z}}$ is simplified to estimating the vector function $F_{P_{X|Z}} : \mathcal{Z} \to \mathcal{H}_\mathcal{X}$. This is equivalent to vector-valued regression with input space $\mathcal{Z}$ and output space $\mathcal{H}_\mathcal{X}$, so that the problem of estimating $F_{P_{X|Z}}$ can be reformulated as the following optimisation:

$$\hat{F}_{P_{X|Z}} = \underset{F \in \mathcal{G}_{\mathcal{X}\mathcal{Z}}}{\text{argmin}} \; \mathcal{E}_{X|Z}(F), \tag{11}$$

where the loss is given by

$$\mathcal{E}_{X|Z}(F) := E_Z\left[||F_{P_{X|Z}}(Z) - F(Z)||^2_{\mathcal{H}_\mathcal{X}}\right], \tag{12}$$

and $\mathcal{G}_{\mathcal{X}\mathcal{Z}}$ is a vector-valued RKHS of functions $\mathcal{Z} \to \mathcal{H}_\mathcal{X}$ with kernel $\ell_{\mathcal{X}\mathcal{Z}}(z,z') = k_\mathcal{Z}(z,z')$ **Id**, where $k_\mathcal{Z}(\cdot,\cdot)$ is a scalar kernel on $\mathcal{Z}$ and **Id** is the identity operator.

The minimisation of $\mathcal{E}_{X|Z}$ cannot be carried out directly, since we only observe pairs (x_i, z_i) from (X, Z). Therefore, we proceed in two steps (see [8] for details). First, we bound our target loss, $\mathcal{E}_{X|Z}$, with a surrogate loss, $\tilde{\mathcal{E}}_{X|Z}$:

$$
\begin{aligned}
\mathcal{E}_{X|Z}(F) &= E_Z \left[\|E_{X|Z}\left[k_{\mathcal{X}}(X, \cdot) - F(Z)|Z\right]\|^2_{\mathcal{H}_{\mathcal{X}}}\right] \\
&\leq E_Z \left[E_{X|Z}\left[\|k_{\mathcal{X}}(X, \cdot) - F(Z)\|^2_{\mathcal{H}_{\mathcal{X}}}|Z\right]\right] \\
&= E_{X,Z} \left[\|k_{\mathcal{X}}(X, \cdot) - F(Z)\|^2_{\mathcal{H}_{\mathcal{X}}}\right] =: \tilde{\mathcal{E}}_{X|Z}(F)
\end{aligned}
$$

Second, we use theoretical results to estimate empirically the surrogate loss $\tilde{\mathcal{E}}_{X|Z}$ from sample data, by using a regularised loss function:

$$
\tilde{\mathcal{E}}_{X|Z,N,\lambda}(F) := \frac{1}{N}\sum_{n=1}^{N} \|k_{\mathcal{X}}(x^{(n)}, \cdot) - F(z^{(n)})\|^2_{\mathcal{H}_{\mathcal{X}}} + \lambda\|F\|^2_{\mathcal{G}_{\mathcal{X}\mathcal{Z}}}, \qquad (13)
$$

where λ is a regularisation parameter and the norm is computed using the kernel $\ell(\cdot, \cdot)$. Given samples $\{(x^{(n)}, z^{(n)})\}_{n=1}^{N}$ from the joint distribution P_{XZ}, our estimation problem is thus relaxed to the minimisation of the population version of the surrogate loss optimisation (13):

$$
\hat{F}_{P_{X|Z,N,\lambda}} = \underset{F \in \mathcal{G}_{\mathcal{X}\mathcal{Z}}}{\text{argmin}} \ \tilde{\mathcal{E}}_{X|Z,N,\lambda}(F). \qquad (14)
$$

The solution to this problem is fully characterised by a Theorem by [7] (see Appendix B). Let us define $N \times 1$ column vector functions $\mathbf{k}_X(\cdot) := (k_{\mathcal{X}}(x^{(1)}, \cdot), \ldots, k_{\mathcal{X}}(x^{(N)}, \cdot))^T$ and $\mathbf{k}_Z(\cdot) := (k_{\mathcal{Z}}(z^{(1)}, \cdot), \ldots, k_{\mathcal{Z}}(z^{(N)}, \cdot))^T$, and the $N \times N$ kernel matrix $\mathbf{K}_Z$ with elements $[\mathbf{K}_Z]_{ij} := k_{\mathcal{Z}}(z^{(i)}, z^{(j)})$. First, we obtain the solutions $\mathbf{u} := (u^{(1)}, \ldots, u^{(N)})^T$, $u^{(i)} \in \mathcal{H}_{\mathcal{X}}$ of the linear system

$$
(\mathbf{K}_Z + N\lambda\mathbf{I})\,\mathbf{u} = \mathbf{k}_X,
$$

where $\mathbf{I}$ is the $N \times N$ identity matrix. Then the solution of the minimisation (14), if it exists, is unique up to a set of measure zero and is given by:

$$
\hat{F}_{P_{X|Z},N,\lambda}(\cdot) = \mathbf{k}_Z^T(\cdot)\mathbf{W}_Z\,\mathbf{k}_X \in \mathcal{G}_{\mathcal{X}\mathcal{Z}},
$$

where $\mathbf{W}_Z := (\mathbf{K}_Z + N\lambda\mathbf{I})^{-1}$. With these results in place, we can construct the empirical estimator of the MCMD between the distributions $P_{X|Z}$ and $P_{X'|Z'}$.

Definition 10 (Estimator of the MCMD). *Let $\{(x^{(n)}, z^{(n)})\}_{n=1}^{N}$ and $\{(x'^{(n)}, z'^{(n)})\}_{n=1}^{N}$ be samples from distributions P_{XZ} and $P_{X'Z'}$, respectively. Consider $N \times N$ kernel matrices $\mathbf{K}_X$, $\mathbf{K}_{X'}$, $\mathbf{K}_{XX'}$, $\mathbf{K}_Z$ and $\mathbf{K}_{Z'}$ with elements $[\mathbf{K}_X]_{st} = k_{\mathcal{X}}(x^{(s)}, x^{(t)})$, $[\mathbf{K}_{XX'}]_{st} = k_{\mathcal{X}}(x^{(s)}, x'^{(t)})$, and similarly for the others. Let us also define $N \times 1$ column vector functions $\mathbf{k}_Z(\cdot) = (k_{\mathcal{Z}}(z^{(1)}, \cdot), \ldots, k_{\mathcal{Z}}(z^{(N)}, \cdot))^T$ and $\mathbf{k}_{Z'}(\cdot) = (k_{\mathcal{Z}}(z'^{(1)}, \cdot), \ldots, k_{\mathcal{Z}}(z'^{(N)}, \cdot))^T$.*

Then the estimator of the MCMD is defined as

$$
\begin{aligned}
\widehat{MCMD}_{P_{X|Z},P_{X'|Z'}}(\cdot) &= \|\hat{F}_{P_{X|Z},N,\lambda}(\cdot) - \hat{F}_{P_{X'|Z'},N,\lambda}(\cdot)\|_{\mathcal{H}_{\mathcal{X}}} \\
&= \big[\mathbf{k}_Z^T(\cdot)\mathbf{W}_Z\mathbf{K}_X\mathbf{W}_Z\,\mathbf{k}_Z(\cdot) + \mathbf{k}_{Z'}^T(\cdot)\mathbf{W}_{Z'}\mathbf{K}_{X'}\mathbf{W}_{Z'}\,\mathbf{k}_{Z'}(\cdot) \\
&\quad - 2\,\mathbf{k}_Z^T(\cdot)\mathbf{W}_Z\mathbf{K}_{XX'}\mathbf{W}_{Z'}\,\mathbf{k}_{Z'}(\cdot)\big]^{1/2},
\end{aligned}
$$
$$
(15)
$$

where $\mathbf{W}_Z = (\mathbf{K}_Z + N\lambda\,\mathbf{I})^{-1}$ *and* $\mathbf{W}_{Z'} = (\mathbf{K}_{Z'} + N\lambda\,\mathbf{I})^{-1}$.

3   Interventional Mean Embeddings and Interventional MCMD

The RKHS machinery developed for conditional mean embeddings and the estimation of the maximum conditional mean discrepancy can be applied to interventional distributions $P_{X_j|do(X_i)=\hat{x}_i}$, where X_i is the intervened node, X_j is the target node, and $\mathbf{Z}$ is a valid adjustment set for the ordered pair (X_i, X_j).

Let $X_d : \Omega \rightarrow \mathcal{X}_d, 1 \leq d \leq D$ be random variables where $(\mathcal{X}_d, \mathfrak{X}_d)$ are separable measurable spaces, and $\mathcal{H}_{\mathcal{X}_d}$ is the RKHS of functions on $\mathcal{X}_d$ with reproducing kernel $k_{\mathcal{X}_d}(\cdot, \cdot)$. Accordingly, given that $\mathbf{Z}$ is a subset of the random variables conforming the valid adjustment set, let $\mathbf{Z} : \Omega \rightarrow \mathcal{Z}$ where $\mathcal{Z}$ is the product space of $\mathcal{X}_d$'s and $\mathfrak{Z}$ is the product σ-algebra of the $\mathfrak{X}_d$'s corresponding to the random variables in $\mathbf{Z}$.

First, note that from Theorem 2, it follows that

$$\mu_{P_{X_j|X_i,\mathbf{Z}}} = \mathbb{E}_{X_j|X_i,\mathbf{Z}}[k_{\mathcal{X}_j}(X_j, \cdot)|X_i, \mathbf{Z}] = F_{P_{X_j|X_i,\mathbf{Z}}} \circ (X_i, \mathbf{Z}), \qquad (16)$$

where $F_{P_{X_j|X_i,\mathbf{Z}}} : \mathcal{X}_i \times \mathcal{Z} \rightarrow \mathcal{H}_{\mathcal{X}_j}$ is a deterministic function measurable with respect to $\mathfrak{X}_i \times \mathfrak{Z}$ and $\mathcal{B}(\mathcal{H}_{\mathcal{X}_j})$.

Definition 11 (Interventional mean embedding).
The mean embedding of $P_{X_j|do(X_i)=\hat{x}_i}$, denoted the interventional mean embedding (IME), is given by

$$\mu_{P_{X_j|do(X_i)=\hat{x}_i}} := \int_{\mathcal{X}_j} k_{\mathcal{X}_j}(x_j, \cdot)\, p\,(x_j|do(X_i) = \hat{x}_i)\, dx_j\,, \qquad (17)$$

which follows from the general definition (7).

Using the definition of valid adjustment set in Eq. (4) and Theorem 2, we get

$$\mu_{P_{X_j|do(X_i)=\hat{x}_i}} = \int_{\mathcal{X}_j} k_{\mathcal{X}_j}(x_j, \cdot) \left(\int_{\mathcal{Z}} p(x_j|\hat{x}_i, \mathbf{z})\, p(\mathbf{z})d\mathbf{z} \right) dx_j \qquad (18)$$

$$= \int_{\mathcal{Z}} \left(\int_{\mathcal{X}_j} k_{\mathcal{X}_j}(x_j, \cdot)\, p(x_j|\hat{x}_i, \mathbf{z})dx_j \right) p(\mathbf{z})d\mathbf{z} \qquad (19)$$

$$= \int_{\mathcal{Z}} F_{P_{X_j|X_i,\mathbf{Z}}}(\hat{x}_i, \mathbf{z})\, p(\mathbf{z})d\mathbf{z} = \mathbb{E}_{\mathbf{Z}}\left[F_{P_{X_j|X_i,\mathbf{Z}}}(\hat{x}_i, \mathbf{Z}) \right]. \qquad (20)$$

Remark 5. Note that the interventional distribution $P_{X_j|do(X_i)=\hat{x}_i}$ remains unchanged for any choice of a valid adjustment set, as shown by Eq. (4). Therefore, it follows that the IME (17) is invariant to the choice of the valid adjustment set, $\mathbf{Z}$, for a given DAG.

With these results, we can define the MCMD between the interventional distributions estimated using different subsets.

Definition 12 (Plug-in function for the family of IMEs). *The family of mean embeddings of interventional distribution $P_{X_j|do(X_i)}$ is given by:*

$$G_{P_{X_j|do(X_i)}}(\cdot) = \mathbb{E}_{\mathbf{Z}}[F_{P_{X_j|X_i,\mathbf{z}}}(\cdot, \mathbf{Z})], \tag{21}$$

where $G_{P_{X_j|do(X_i)}} : \mathcal{X}_i \to \mathcal{H}_{\mathcal{X}_j}$ is the measurable, deterministic functions that maps each possible intervention $\hat{x}_i \in \mathcal{X}_i$ to the embedding of its interventional distribution $P_{X_j|do(X_i)=\hat{x}_i}$.

Definition 13 (MCMD between estimated interventional distributions). *Let $P_{X_j|do(X_i)}$ and $P'_{X_j|do(X_i)}$ be the interventional distributions estimated using two different subsets $\mathbf{Z}$ and $\mathbf{Z}'$, with plug-in embedding functions $G_{P_{X_j|do(X_i)}}$ and $G_{P'_{X_j|do(X_i)}}$, respectively, defined as in (21). Then the MCMD between these interventional distributions is*

$$MCMD_{P_{X_j|do(X_i)},P'_{X_j|do(X_i)}}(\cdot) = ||G_{P_{X_j|do(X_i)}}(\cdot) - G_{P'_{X_j|do(X_i)}}(\cdot)||_{\mathcal{H}_{\mathcal{X}_j}}, \tag{22}$$

where $MCMD_{P_{X_j|do(X_i)},P'_{X_j|do(X_i)}}(\cdot) : \mathcal{X}_i \to \mathbb{R}$.

Next, we derive an empirical estimate of the interventional MCMD (22). We follow closely Sect. 2.2 but instead of conditioning only on one variable, as in Eq. (8), here we condition on the intervened variable, X_i, and a subset, $\mathbf{Z}$. This is achieved by considering product kernels.

Similarly to (11)–(12), our goal is to minimise the loss function

$$\mathcal{E}_{X_j|X_i,\mathbf{Z}}(F) = \mathbb{E}_{X_i,\mathbf{Z}}\left[||F_{P_{X_j|X_i,\mathbf{z}}}(X_i, \mathbf{Z}) - F(X_i, \mathbf{Z})||^2_{\mathcal{H}_{\mathcal{X}_j}}\right] \tag{23}$$

over all vector functions $F : \mathcal{X}_i \times \mathbf{Z} \to \mathcal{H}_{\mathcal{X}_j}$ that form an RKHS $\mathcal{G}_{\mathcal{X}_j,\mathcal{X}_i\mathbf{Z}}$ with kernel $l_{\mathcal{X}_j,\mathcal{X}_i\mathbf{Z}}((x_i, \mathbf{z}), (x'_i, \mathbf{z}')) = k_{\mathcal{X}_i\mathbf{Z}}((x_i, \mathbf{z}), (x'_i, \mathbf{z}')) \mathbf{Id}$, where $\mathbf{Id}$ is the identity operator and $k_{\mathcal{X}_i\mathbf{Z}}$ is the product kernel on $\mathcal{X}_i \times \mathbf{Z}$ given by

$$k_{\mathcal{X}_i\mathbf{Z}}((x_i, \mathbf{z}), (x'_i, \mathbf{z}')) = k_{\mathcal{X}_i}(x_i, x'_i)k_{\mathbf{Z}}(\mathbf{z}, \mathbf{z}') = k_{\mathcal{X}_i}(x_i, x'_i) \prod_{\alpha=1}^{M} k_{\mathcal{X}_\alpha}(x_\alpha, x'_\alpha) \tag{24}$$

where $\mathbf{Z}$ is a subset of M variables, i.e., $\mathbf{Z} = \{X_\alpha\}_{\alpha=1}^{M}$. The optimisation of $\mathcal{E}_{X_j|X_i,\mathbf{Z}}(F)$ is relaxed following the same steps detailed in Sect. 2.2:

– First, we produce a lower bound for $\mathcal{E}_{X_j|X_i,\mathbf{Z}}(F)$ which we use as a surrogate loss:

$$\mathcal{E}_{X_j|X_i,\mathbf{Z}}(F) = \mathbb{E}_{X_i,\mathbf{Z}}\left[||\mathbb{E}_{X_j|X_i,\mathbf{Z}}\left[k_{\mathcal{X}_j}(X_j, \cdot) - F(X_i, \mathbf{Z})\right]|X_i, \mathbf{Z}||^2_{\mathcal{H}_{\mathcal{X}_j}}\right]$$

$$\leq \mathbb{E}_{X_i,\mathbf{Z}}\left[\mathbb{E}_{X_j|X_i,\mathbf{Z}}\left[||k_{\mathcal{X}_j}(X_j, \cdot) - F(X_i, \mathbf{Z})||^2_{\mathcal{H}_{\mathcal{X}_j}}|X_i, \mathbf{Z}\right]\right]$$

$$= \mathbb{E}_{X_i,X_j,\mathbf{Z}}\left[||k_{\mathcal{X}_j}(X_j, \cdot) - F(X_i, \mathbf{Z})||^2_{\mathcal{H}_{\mathcal{X}_j}}\right] =: \tilde{\mathcal{E}}_{X_j|X_i,\mathbf{Z}}(F)$$

- Second, we introduce the empirical regularised version of the surrogate loss function:

$$\hat{\mathcal{E}}_{X_j|X_i,\mathbf{Z},N,\lambda}(F) := \frac{1}{N}\sum_{n=1}^{N}||k_{\mathcal{X}_j}(x_j^{(n)},\cdot) - F(x_i^{(n)},\mathbf{z}^{(n)})||_{\mathcal{H}_{\mathcal{X}_j}}^2 + \lambda\,||F||_{\mathcal{G}_{\mathcal{X}_j,\mathcal{X}_i\mathbf{Z}}}^2$$

where $\{x_i^{(n)}, x_j^{(n)}, \mathbf{z}^{(n)}\}_{n=1}^N$ are samples from the joint distribution $P_{X_iX_j\mathbf{Z}}$.
- Third, we use Theorem 3 to find the unique minimiser, up to a set of measure zero, of $\hat{\mathcal{E}}_{X_j|X_i,\mathbf{Z},N,\lambda}(F)$:

$$\hat{F}_{P_{X_j|X_i,\mathbf{Z},N,\lambda}}(\cdot,\cdot) = \mathbf{k}_{X_i\mathbf{Z}}(\cdot,\cdot)^T\mathbf{W}\,\mathbf{k}_{X_j}. \tag{25}$$

Here $\mathbf{k}_{X_i\mathbf{Z}}(\cdot,\cdot)$ and $\mathbf{k}_{X_j}$ are kernel column vector functions, and $\mathbf{W} = (\mathbf{K}_{X_i\mathbf{Z}} + N\lambda\mathbf{I})^{-1}$, where the $N \times N$ kernel matrix $\mathbf{K}_{X_i\mathbf{Z}}$ has elements $[\mathbf{K}_{X_i\mathbf{Z}}]_{st} := k_{\mathcal{X}_i\mathbf{Z}}((x_i^{(s)},\mathbf{z}^{(s)}),(x^{(t)},\mathbf{z}^{(t)}))$.
- Fourth, we obtain the empirical estimate for $G_{P_{X_j|do(X_i)}} : \mathcal{X}_i \to \mathcal{H}_{\mathcal{X}_j}$.

$$\hat{G}_{P_{X_j|do(X_i)}}(\cdot) = \frac{1}{N}\sum_{n=1}^{N}\mathbf{k}_{X_i\mathbf{Z}}^T(\cdot,\mathbf{z}^{(n)})\mathbf{W}\,\mathbf{k}_{X_j}$$

- Finally, given samples $\{(x_i^{(n)}, x_j^{(n)}, z^{(n)})\}_{n=1}^N$ and $\{(x_i^{(n)}, x_j^{(n)}, z'^{(n)})\}_{n=1}^N$ from $P_{X_iX_jZ}$ and $P_{X_iX_jZ'}$, we use the definition (22) to obtain an empirical estimate for the MCMD between two interventional distributions $P_{X_j|do(X_i)}$ and $P'_{X_j|do(X_i)}$ estimated using two different subsets $\mathbf{Z}$ and $\mathbf{Z}'$:

$$\widehat{MCMD}_{P_{X_j|do(X_i)},P'_{X_j|do(X_i)}}(\cdot) = ||\hat{G}_{P_{X_j|do(X_i)}}(\cdot) - \hat{G}_{P'_{X_j|do(X_i)}}(\cdot)||_{\mathcal{H}_{\mathcal{X}_j}}$$

$$= \left[\left(\frac{1}{N}\sum_{n=1}^{N}\mathbf{k}_{X_i\mathbf{Z}}^T(\cdot,\mathbf{z}^{(n)})\right)\mathbf{W}_{\mathbf{Z}}\mathbf{K}_{X_j}\mathbf{W}_{\mathbf{Z}}\left(\frac{1}{N}\sum_{n=1}^{N}\mathbf{k}_{X_i\mathbf{Z}}(\cdot,\mathbf{z}^{(n)})\right)\right.$$

$$+ \left(\frac{1}{N}\sum_{n=1}^{N}\mathbf{k}_{X_i\mathbf{Z}'}^T(\cdot,\mathbf{z}'^{(n)})\right)\mathbf{W}_{\mathbf{Z}'}\mathbf{K}_{X_j}\mathbf{W}_{\mathbf{Z}'}\left(\frac{1}{N}\sum_{n=1}^{N}\mathbf{k}_{X_i\mathbf{Z}'}(\cdot,\mathbf{z}'^{(n)})\right)$$

$$\left.-2\left(\frac{1}{N}\sum_{n=1}^{N}\mathbf{k}_{X_i\mathbf{Z}}^T(\cdot,\mathbf{z}^{(n)})\right)\mathbf{W}_{\mathbf{Z}}\mathbf{K}_{X_j}\mathbf{W}_{\mathbf{Z}'}\left(\frac{1}{N}\sum_{n=1}^{N}\mathbf{k}_{X_i\mathbf{Z}'}(\cdot,\mathbf{z}'^{(n)})\right)\right]^{1/2}$$

where $\mathbf{W}_{\mathbf{Z}'} = (\mathbf{K}_{X_i\mathbf{Z}'} + N\lambda\mathbf{I})^{-1}$ with $[\mathbf{K}_{X_i\mathbf{Z}'}]_{st} := k_{\mathcal{X}_i\mathbf{Z}}((x_i^{(s)},\mathbf{z}'^{(s)}),(x^{(t)},\mathbf{z}'^{(t)}))$, and the other kernel matrices and kernel vector functions are defined similarly to above.

4   Continuous Structural Intervention Distance

Consider the setting where we have a true DAG $\mathcal{G}_1 = (\mathbf{V},\mathcal{E}_{\mathcal{G}_1})$, a learnt DAG $\mathcal{G}_2 = (\mathbf{V},\mathcal{E}_{\mathcal{G}_2})$ and observational data $\mathcal{D}$ sampled from an unknown distribution P with density $p(\cdot)$ that is Markov with respect to $\mathcal{G}_1$. Note that the true and learnt DAGs have a common set of vertices but differ in their edges. To define the true interventional distribution, we consider an additional D-dimensional joint distribution $\widehat{P}$ and samples $\widehat{\mathcal{D}} = \{\hat{x}_1^{(m)}, \ldots, \hat{x}_D^{(m)}\}_{m=1}^M$ drawn from $\widehat{P}$, which resemble the joint distribution of the variables $X_1, \ldots, X_D$ after we intervened on them (see Remark 7). Let $\widehat{P}_i$ denote the marginal distribution of $\widehat{P}$ in the ith argument. Then the true interventional distribution, $P_{X_j|do(X_i)=\hat{x}_i}$ where $\hat{x}_i \sim \widehat{P}_i$, has density $\mathbb{E}_{\hat{X}_i}[p(X_j|do(X_i) = \hat{x}_i)]$. If $\tilde{\mu}_{P_{X_j|do(X_i)=\hat{x}_i}}$ is the empirical estimate of the mean embedding of the distribution $P_{X_j|do(X_i)=\hat{x}_i}$, then the empirical estimate of the mean embedding of the true interventional distribution is $\tilde{\mu}_{P_{X_j|do(X_i)=\hat{x}_i}} = \frac{1}{M}\sum_{m=1}^M \tilde{\mu}_{P_{X_j|do(X_i)=\hat{x}_i^{(m)}}}$. Let $P_{X_j|do(X_i)=\hat{x}_i;\mathcal{G}_1}$ and $P_{X_j|do(X_i)=\hat{x}_i;\mathcal{G}_2}$ denote the estimate of $P_{X_j|do(X_i)=\hat{x}_i}$ with respect to the true DAG $\mathcal{G}_1$ and the learnt DAG $\mathcal{G}_2$, respectively. In other words, $P_{X_j|do(X_i)=\hat{x}_i;\mathcal{G}_k}$ is our estimate of the true interventional distribution given $\mathcal{G}_k, \mathcal{D}$ and $\widehat{\mathcal{D}}$. Since $\mathcal{G}_1$ is the true DAG, and P is Markov with respect to $\mathcal{G}_1$, $P_{X_j|do(X_i);\mathcal{G}_1}$ is exactly equal to the true interventional distribution $P_{X_j|do(X_i)=\hat{x}_i}$. First, we generate the set $\mathbf{V}^2 := (\mathbf{V} \times \mathbf{V})$, which consists of all ordered pairs of nodes from the common vertex set of the true DAG and the learnt DAG. For each pair $(X_i, X_j) \in \mathbf{V}^2, i \neq j$, we compare $P_{X_j|do(X_i)=\hat{x}_i;\mathcal{G}_1}$ and $P_{X_j|do(X_i)=\hat{x}_i;\mathcal{G}_2}$ (this can be extended to multiple simultaneous interventions—see Remark 6). We record the norm difference of the embeddings of $P_{X_j|do(X_i)=\hat{x}_i;\mathcal{G}_1}$ and $P_{X_j|do(X_i)=\hat{x}_i;\mathcal{G}_2}$ in the function $d : \tilde{\mathbf{V}}^2 \to \mathbb{R}_{\geq 0}$ which we describe below by examining various possible cases of edges that can occur in the graphs $\mathcal{G}_1$ and $\mathcal{G}_2$.

Case 1: There is no directed path from X_i to X_j in DAGs $\mathcal{G}_1$ and $\mathcal{G}_2$ (in Algorithm 1 this condition is checked by the function "checkDirectedPath$(X_i, X_j, \mathcal{G})$"). In the absence of a directed path from the intervened node to the target node, an intervention has no effect on the target node. So, in $\mathcal{G}_1$ and $\mathcal{G}_2$ the distribution of X_j obtained by intervening on X_i is equal to the observational distribution of X_j, i.e., $P_{X_j|do(X_i)=\hat{x}_i;\mathcal{G}_1} = P_{X_j|do(X_i)=\hat{x}_i;\mathcal{G}_2} = P_{X_j}$. This in turn implies $d(X_i, X_j) = 0$.

Case 2: There is a directed path from X_i to X_j in $\mathcal{G}_1$ but not in $\mathcal{G}_2$. The same argument used in Case 1 can be applied here to obtain $P_{X_j|do(X_i)=\hat{x}_i;\mathcal{G}_2} = P_{X_j}$. Intervening on X_i has an effect on X_j in $\mathcal{G}_1$ due to the presence of the directed path $X_i \to X_j$ and so, $P_{X_j|do(X_i)=\hat{x}_i;\mathcal{G}_1}$ can be computed by adjusting for the parent set. We compare the two distributions $P_{X_j|do(X_i)=\hat{x}_i;\mathcal{G}_1}$ and P_{X_j} by computing the average over their MMDs for each sample $\hat{x}_i$ of the distribution $\widehat{P}_i$. We then divide by the norm of the embedding of the observational distribution of X_j to make contSID scale-invariant. The resulting distance d is defined as we state in Eq. (26), where we denote $\frac{1}{N}\left(\sum_{n,n'=1}^N k_{X_j}(x_j^{(n)}, x_j^{(n')})\right)^{1/2}$ by C_{X_j} and $\frac{1}{\sqrt{MN}}\left(\sum_{n=1}^N \sum_{m=1}^M \mathbf{k}_{X_i\mathbf{PA}_{i,\mathcal{G}_1}}(\hat{x}_i^{(m)}, \mathbf{pa}_{i,\mathcal{G}_1}^{(n)})\right)^{1/2}$ by $\mathbf{k}_{X_i,\mathcal{G}_1}$ ($\mathbf{k}_{X_i,\mathcal{G}_2}$ is analogously defined).

$$d(X_i, X_j) = \frac{1}{C_{X_j}} \| \tilde{\mu}_{P_{X_j | do(X_i); \mathcal{G}_1}} - \tilde{\mu}_{P_{X_j}} \|_{\mathcal{H}_{X_j}}$$

$$= \frac{1}{C_{X_j}} \Big[\mathbf{k}_{X_i, \mathcal{G}_1}^T \mathbf{W}_{\mathcal{G}_1} \mathbf{K}_{X_j} \mathbf{W}_{\mathcal{G}_1} \mathbf{k}_{X_i, \mathcal{G}_1} + C_{X_j}^2 \tag{26}$$

$$-2\mathbf{k}_{X_i, \mathcal{G}_1}^T \mathbf{W}_{\mathcal{G}_1} \left(\frac{1}{N} \sum_{n=1}^{N} \mathbf{k}_{X_j} (x_j^{(n)}) \right) \Big]^{1/2}$$

Similarly, if there is a directed path from X_i to X_j in $\mathcal{G}_2$ but not in $\mathcal{G}_1$, the resulting distance d is:

$$d(X_i, X_j) = \frac{1}{C_{X_j}} \Big[\mathbf{k}_{X_i, \mathcal{G}_2}^T \mathbf{W}_{\mathcal{G}_2} \mathbf{K}_{X_j} \mathbf{W}_{\mathcal{G}_2} \mathbf{k}_{X_i, \mathcal{G}_2} + C_{X_j}^2$$

$$-2\mathbf{k}_{X_i, \mathcal{G}_2}^T \mathbf{W}_{\mathcal{G}_2} \left(\frac{1}{N} \sum_{n=1}^{N} \mathbf{k}_{X_j} (x_j^{(n)}) \right) \Big]^{1/2} \tag{27}$$

Case 3: There is a directed path from X_i to X_j in DAG $\mathcal{G}_1$ and $\mathcal{G}_2$. The distribution of X_j after intervening on X_i in $\mathcal{G}_1$ can be computed by adjusting for the parent set of X_i in $\mathcal{G}_1$, $\mathbf{PA}_{i;\mathcal{G}_1}$. Similarly, we obtain the interventional distribution of X_j in $\mathcal{G}_2$ by adjusting for the parent set of X_i in $\mathcal{G}_2$, $\mathbf{PA}_{i;\mathcal{G}_2}$.

1. If $\mathbf{PA}_{i;\mathcal{G}_1}$ is a valid adjustment set (Definition 3) in $\mathcal{G}_2$ or $\mathbf{PA}_{i;\mathcal{G}_2}$ is a valid adjustment set in $\mathcal{G}_1$, then by Eq. (4), $P_{X_j | do(X_i); \mathcal{G}_1} = P_{X_j | do(X_i); \mathcal{G}_2}$, hence $d(X_i, X_j) = 0$.[1]
2. If $\mathbf{PA}_{i;\mathcal{G}_1}$ is not a valid adjustment set in $\mathcal{G}_2$ or $\mathbf{PA}_{i;\mathcal{G}_2}$ is not a valid adjustment set in $\mathcal{G}_1$, then the interventional distributions $P_{X_j | do(X_i); \mathcal{G}_1}$ and $P_{X_j | do(X_i); \mathcal{G}_2}$ *may not* be equal. To assess the difference, we compute the average over their MMDs for each $x_i \sim \mathcal{D}_i$. We divide by the norm of the embedding of the observational distribution X_j to make contSID scale-invariant. The resulting distance d is defined as we state in Eq. (28).

$$d(X_i, X_j) = \frac{1}{C_{X_j}} \| \tilde{\mu}_{P_{X_j | do(X_i); \mathcal{G}_2}} - \tilde{\mu}_{P_{X_j | do(X_i); \mathcal{G}_1}} \|_{\mathcal{H}_{X_j}}$$

$$= \frac{1}{C_{X_j}} \Big[\mathbf{k}_{X_i, \mathcal{G}_2} \mathbf{W}_{\mathcal{G}_2} \mathbf{K}_{X_j} \mathbf{W}_{\mathcal{G}_2} \mathbf{k}_{X_i, \mathcal{G}_2} + \mathbf{k}_{X_i, \mathcal{G}_1} \mathbf{W}_{\mathcal{G}_1} \mathbf{K}_{X_j} \mathbf{W}_{\mathcal{G}_1} \mathbf{k}_{X_i, \mathcal{G}_1}$$

$$-2\mathbf{k}_{X_i, \mathcal{G}_2} \mathbf{W}_{\mathcal{G}_2} \mathbf{K}_{X_j} \mathbf{W}_{\mathcal{G}_1} \mathbf{k}_{X_i, \mathcal{G}_1} \Big]^{1/2}$$

$$\tag{28}$$

We summarise the various cases and the applicable equations in Algorithm 1, and describe how the contSID is calculated over each ordered pair $(X_i, X_j) \in \mathbf{V}^2, i \neq j$ in Algorithm 2.

---

[1] In general, the above condition is not necessary for $P_{X_j | do(X_i); \mathcal{G}_1} = P_{X_j | do(X_i); \mathcal{G}_2}$. It is sufficient that there is a common valid adjustment set—not just a parent adjustment set—for the pair (X_i, X_j) in $\mathcal{G}_1$ and $\mathcal{G}_2$. However, it is not straightforward and beyond the scope of this article to compare the validity of an adjustment in different DAGs. Thus, we resort to the simple and inexpensive graphical task of checking if the parent sets in one DAG are valid adjustment sets in the other DAG.

Remark 6 (Interventions on multiple variables). As in [10], we have considered intervening on single variables only for simplification purposes here. However, the contSID can be extended to account for interventions on multiple variables as well. Since the union of parent sets of the intervened variables is not necessarily a valid adjustment set, one would need to define a valid adjustment set for the intervened variables and the observed variable. Then, using a modified version of Eq. (4), we can compute the interventional distribution and its corresponding embedding. This can be achieved by replacing the one intervened variable $X_i :$ $\Omega \to \mathcal{X}$ with the set of variables $\mathbf{X}_i : \Omega \to \boldsymbol{\mathcal{X}}_i$ that we intervene on, and defining the corresponding kernel $k_{\boldsymbol{\mathcal{X}}_i} : \boldsymbol{\mathcal{X}}_i \times \boldsymbol{\mathcal{X}}_i \to \mathbb{R}$.

Remark 7 (Distribution on interventions). Unless specified otherwise, we use the empirical distribution of X_i to compute the average of the MMDs in Eqs. (26), (27) and (28) to determine the contSID. If required, however, one may specify an alternative distribution on the intervention, e.g., assigning a Dirac measure as a single, "hard" intervention, and evaluating the contSID with that distribution.

Algorithm 1: $d(X_i, X_j, \mathcal{G}_1, \mathcal{G}_2, \mathcal{D})$

Input: Intervened node X_i, target node X_j, true DAG $\mathcal{G}_1 = (\mathbf{V}, E_{\mathcal{G}_1})$, learnt DAG $\mathcal{G}_2 = (\mathbf{V}, E_{\mathcal{G}_2})$ and the observational data $\mathcal{D}$

$c_{\mathcal{G}_1} \leftarrow \text{checkDirectedPath}(X_i, X_j, \mathcal{G}_1)$
$c_{\mathcal{G}_2} \leftarrow \text{checkDirectedPath}(X_i, X_j, \mathcal{G}_2)$

if $c_{\mathcal{G}_1} ==$ *False and* $c_{\mathcal{G}_2} ==$ *False* **then**
  | **return** 0
else
  | $Z_{\mathcal{G}_1} \leftarrow \mathbf{PA}_{i,\mathcal{G}_1}$ $Z_{\mathcal{G}_2} \leftarrow \mathbf{PA}_{i,\mathcal{G}_2}$
end

$K \leftarrow \sum_{m,m'}^{N} k(x_j^{(m)}, x_j^{(m')})$

if $c_{\mathcal{G}_1} ==$ *True and* $c_{\mathcal{G}_2} ==$ *False* **then**
  | **return** (26)
else if $c_{\mathcal{G}_1} ==$ *False and* $c_{\mathcal{G}_2} ==$ *True* **then**
  | **return** (27)
else
  | **if** $Z_{\mathcal{G}_1}$ *is a valid adjustment set in* $\mathcal{G}_2$ *or* $Z_{\mathcal{G}_2}$ *is a valid adjustment set in* $\mathcal{G}_1$
  | **then**
  | | **return** 0
  | **else**
  | | **return** (28)
  | **end**
end

Algorithm 2: contSID($\mathcal{G}_1, \mathcal{G}_2, \mathcal{D}$)

Input: True DAG $\mathcal{G}_1 = (\mathbf{V}, E_{\mathcal{G}_1})$, learnt DAG $\mathcal{G}_2 = (\mathbf{V}, E_{\mathcal{G}_2})$ and the
observational data $\mathcal{D}$

sum $\leftarrow 0$
for $(X_i, X_j) \in \mathbf{V}^2, \quad i \neq j$ **do**
| sum = sum + $d(X_i, X_j, \mathcal{G}_1, \mathcal{G}_2, \mathcal{D})$
end
return sum

---

5 Experiments

For each number of nodes $p \in \{5, 10, 20\}$, we generate 100 DAGs by an Erdos-Rènyi model with the probability of the existence of an edge equal to 0.25. 100 *iid* samples $\mathcal{D} \in \mathbb{R}^p$ are generated for each DAG according to a linear structural equation model (SEM) with non-Gaussian (exponential) noise. The linear coefficients of the SEM are sampled uniformly from the interval $[-10, 10]$, which correspond to edge weights in the interval $[0, 10]$. The exponential noise has scale $\beta = 1$. Then, we conduct two separate experiments. First, for each true, data-generating DAG $\mathcal{G}_1$, we obtain two simulated DAGs that may be outcomes of two different causal discovery algorithms [e.g., 6,13] on the synthetically generated data. These simulated DAGs, $\mathcal{G}_2$ and $\mathcal{G}_3$, are copies of $\mathcal{G}_1$ with the following important modifications. In $\mathcal{G}_2$, the edge with the smallest weight is deleted and in $\mathcal{G}_3$ the edge with the largest weight is deleted. Clearly, $\mathcal{G}_3$ is a much poorer outcome of a learning algorithm than $\mathcal{G}_2$ due to the influence of missing an edge with large weight in subsequent interventions. Accordingly, we may expect that a metric assessing learning algorithms is smaller between $\mathcal{G}_1$ and $\mathcal{G}_2$ than between $\mathcal{G}_1$ and $\mathcal{G}_3$. The results of this experiment can be found in Table 2, which show that contSID is indeed smaller between $\mathcal{G}_1$ and $\mathcal{G}_2$ than between $\mathcal{G}_1$ and $\mathcal{G}_3$ over $p \in \{5, 10, 20\}$. In contrast, SID only behaves according to these expectations for $p = 5$ while it is larger between $\mathcal{G}_1$ and $\mathcal{G}_2$ than between $\mathcal{G}_1$ and $\mathcal{G}_3$ for $p \in \{10, 20\}$.

Table 2. SHD, SID and contSID calculated on $d(\mathcal{G}_1, \mathcal{G}_2)$ and $d(\mathcal{G}_1, \mathcal{G}_3)$ over p. We show averages over 100 randomly generated DAGs plus-minus their standard deviations.

Metric	$d(\mathcal{G}_1, \mathcal{G}_2)$	$d(\mathcal{G}_1, \mathcal{G}_3)$	Metric	$d(\mathcal{G}_1, \mathcal{G}_2)$	$d(\mathcal{G}_1, \mathcal{G}_3)$	Metric	$d(\mathcal{G}_1, \mathcal{G}_2)$	$d(\mathcal{G}_1, \mathcal{G}_3)$
SHD	1.0 ± 0.00	1.0 ± 0.00	SHD	1 ± 0.00	1 ± 0.00	SHD	1 ± 0.00	1 ± 0.00
SID	2.0 ± 0.95	2.08 ± 0.96	SID	5.39 ± 1.65	5.26 ± 1.87	SID	12.54 ± 2.98	12.49 ± 2.77
contSID	0.93 ± 1.01	1.03 ± 0.96	contSID	1.14 ± 1.77	1.16 ± 1.54	contSID	0.88 ± 2.46	0.97 ± 2.15
	(a) $p = 5$			(b) $p = 10$			(c) $p = 20$	

Second, we compare three causal discovery algorithms, namely PC (constraint-based), GES (score-based) and LiNGAM (function-based) [3,13,17, respectively] on synthetically generated data. Here, complete partially directed acyclic graphs (CPDAGs) which entail undirected besides directed edges, and can be the result of the PC and GES algorithms, follow the same cases in Sect. 4 as DAGs (given as an outcome of LiNGAM). Consequently, the criterion "checkDirectedPath($\cdot$)" is false for undirected edges. We compute the average SHD, SID and contSID values as well as their standard deviation for each true and learnt DAG pair. Our results in Table 3a–3c underpin that both SHD and SID only take qualitative graph structures into account, while contSID additionally considers data distributions. When comparing SHD and SID values for the PC, GES and LiNGAM algorithms, we find that LiNGAM strongly outperforms both others, of which GES is slightly more accurate than PC except for $p = 5$ when taking SHD as the accuracy metric. In contrast, contSID tells us that while LiNGAM is still the most accurate algorithm, PC follows second and GES last. The discrepancy between PC and GES is particularly large for $p = 20$. We make code for the implementation of contSID and the experiments available at https://github.com/felix-laumann/contSID.

Table 3. SHD, SID and contSID between learnt and true DAG using different causal discovery algorithms over p. We show averages over 100 randomly generated DAGs plus-minus their standard deviations.

p	PC	GES	LiNGAM
5	2.13 ± 1.32	2.18 ± 1.51	0.89 ± 1.04
10	10.29 ± 3.77	9.67 ± 4.88	3.55 ± 3.34
20	53.1 ± 7.12	47.6 ± 8.87	31.15 ± 10.65

(a) SHD

p	PC	GES	LiNGAM
5	4.7 ± 3.76	4.45 ± 3.82	1.4 ± 2.20
10	37.21 ± 17.65	25.87 ± 14.60	7.86 ± 7.65
20	267.85 ± 39.02	248.23 ± 34.05	124.7 ± 41.50

(b) SID

p	PC	GES	LiNGAM
5	2.43 ± 1.98	2.51 ± 2.11	0.48 ± 0.63
10	20.18 ± 9.35	23.45 ± 12.49	5.28 ± 5.40
20	83.30 ± 37.12	134.37 ± 41.89	51.04 ± 21.60

(c) contSID

6   Conclusion

We propose a novel metric to accurately compare a learnt to a true directed acyclic graph (DAG) in causal structure learning settings. Despite the widespread use of the Structural Hamming Distance (SHD) and the Structural Intervention Distance (SID), two metrics that fulfil the purpose of comparing a learnt to a true DAG, they are based on graph properties only. Besides graph properties, our metric takes additionally the underlying data of the causal system into account and can, hence, distinguish between the importance of learning certain edges more accurately. We argue that this differentiation is especially valuable in graphs with weighted edges. The metric is defined as a distance between kernel conditional mean embeddings that are derived through a measure-theoretic approach. We hope that researchers working on causal structure learning problems find our novel metric useful in their assessment of the accuracy of causal discovery algorithms, and that it can provide additional insights beyond the capabilities of the SHD and SID.

References

1. Acharya, J., Bhattacharyya, A., Daskalakis, C., Kandasamy, S.: Learning and testing causal models with interventions. Advances in Neural Information Processing Systems **31** (2018)
2. Acid, S., de Campos, L.M.: Searching for bayesian network structures in the space of restricted acyclic partially directed graphs. J. Artif. Intell. Res. **18**, 445–490 (2003)
3. Chickering, D.M.: Optimal structure identification with greedy search. J. Mach. Learn. Res. **3**(Nov), 507–554 (2002)
4. Fukumizu, K., Gretton, A., Sun, X., Schölkopf, B.: Kernel measures of conditional dependence. Advances in neural information processing systems **20** (2007)
5. Garant, D., Jensen, D.: Evaluating causal models by comparing interventional distributions. arXiv preprint arXiv:1608.04698 (2016)
6. Hoyer, P., Janzing, D., Mooij, J.M., Peters, J., Schölkopf, B.: Nonlinear causal discovery with additive noise models. Advances in Neural Information Processing Systems (2008)
7. Micchelli, C.A., Pontil, M.: On learning vector-valued functions. Neural Comput. **17**(1), 177–204 (2005)
8. Park, J., Muandet, K.: A measure-theoretic approach to kernel conditional mean embeddings. arXiv preprint arXiv:2002.03689 (2020)
9. Pearl, J.: Causality. Cambridge university press (2009)
10. Peters, J., Bühlmann, P.: Structural intervention distance for evaluating causal graphs. Neural Comput. **27**(3), 771–799 (2015)
11. Peters, J., Janzing, D., Schölkopf, B.: Elements of causal inference. The MIT Press (2017)
12. Peyrard, M., West, R.: A ladder of causal distances. arXiv preprint arXiv:2005.02480 (2020)
13. Shimizu, S., Hoyer, P.O., Hyvärinen, A., Kerminen, A., Jordan, M.: A linear non-gaussian acyclic model for causal discovery. J. Mach. Learn. Res. **7**(10) (2006)
14. Shpitser, I., VanderWeele, T., Robins, J.M.: On the validity of covariate adjustment for estimating causal effects. arXiv preprint arXiv:1203.3515 (2012)
15. Singh, K., Gupta, G., Tewari, V., Shroff, G.: Comparative benchmarking of causal discovery techniques. arXiv preprint arXiv:1708.06246 (2017)
16. Song, L., Huang, J., Smola, A., Fukumizu, K.: Hilbert space embeddings of conditional distributions with applications to dynamical systems. In: Proceedings of the 26th Annual International Conference on Machine Learning, pp. 961–968 (2009)
17. Spirtes, P., Glymour, C.N., Scheines, R.: Causation, prediction, and search. MIT press (2000)
18. Sriperumbudur, B.K., Gretton, A., Fukumizu, K., Schölkopf, B., Lanckriet, G.R.: Hilbert space embeddings and metrics on probability measures. J. Mach. Learn. Res. **11**, 1517–1561 (2010)

Detection Windows from Hidden Markov Model for Discovering Varying Causal Relations Between Time Series

Kaijun Wang, Ying Fang[✉], and Tianjian Luo

College of Computer and Cyber Security, Fujian Normal University, Fuzhou, Fujian 350117,
People's Republic of China
fy20@fjnu.edu.cn

Abstract. The sliding window is commonly used to detect the changing causal relationships between time series, but its performance is sensitive to window sizes. However, little research has been conducted for appropriate window sizes. We propose Detection Windows based on Hidden Markov Model (HMMDW) for time-varying causal discovery of time series. Firstly, a sliding window moves along the time series, an autoregressive model with an external variable and time lags (called full model) are established in each window. A series of adjusted R-squares are calculated by the full model fitted to the time series within each window. Subsequently, these R-squares are as the observations of the hidden Markov model with two states (e.g., 1 & 0), which imply good or poor full models. After the hidden Markov model is solved, the scope of consecutive 1 or 0 corresponds to a detection window size, so a series of detection windows are obtained. Finally, the Granger causality test is used to discover the causal relationships between time series within every detection window. The experimental results show that our method is effective, and its comprehensive performance outperforms the comparative methods in terms of accuracy. Code is available: https://github.com/wkj wang/hmm_window

Keywords: Time Series · Varying Causal Relationships · Granger Causality · Hidden Markov Model

1 Introduction

Exploring causal relationships between time series is importance for understanding the dynamics of evolving systems [1]. The classic approach for discovering causal relationships between bivariate time series is the Granger causality test [2], which means that if adding a variable to a Granger causality model improves its predictive performance, then there exists a statistical causal relationship between the variables in the model. The Granger method is also applicable for studying causal relationships among multivariate time series. There is some related research, for instance, reference [3]. Develops Granger causality methods for anomaly detection of time series, and reference [4]. Proposes a Lasso Granger causality method based on the Hilbert-Schmidt independence criterion for analyzing causal relationships among multivariate time series. The state-space approach [5] is another method for analyzing causal relationships between time series. It describes the time-varying equation parameters by building a state-space model, and identifies causal relationships based on the measurement equation.

For systems that evolve over time, causal relationships between time series (variables) may change over time. We focuses on the topic of causal relationships between two time series changing over time, especially the case that causal relationships appear in one time interval but no causal relationship exists in its adjacent time intervals. Conventional sliding window methods [6, 7]. Employ sliding time windows to search through time series and use Granger causality methods to detect causal relationships between time series within the current window. However, variations in window size and moving-step size may lead to differences in results and performance. The F-bound method [8] Traverses different lengths of time intervals containing time t to detect causal relationships. When a time series region has been detected for causality, in its adjacent time series region the F-bound method uses upper and lower bounds of F-tests to discern causal relationships. The causal ratio, calculated for each t as the ratio of the number of causal intervals containing t to the total number of intervals, serves as a measure of causality between time series at time t. The change point detection method [9]. Identifies change points (as segmentation points) of time series, then performs Granger causality tests in segmented regions to obtain evolving relationships over time. This method assumes that different time intervals separated by segmentation points correspond to changes in causal relationship structures. However, complex time series may interfere the accurate identification of change points, leading to decreased method performance. The different-region balance method [10]. Designs a strategy to detect three distinct regions when the sliding window covers a time series region, then integrates the detection results of these three regions to obtain the causal relationship in the sliding window.

Among these methods, the performance of conventional sliding window methods is sensitive to window sizes and step sizes, leading to instability in performance. The F-test method emphasizes global information by traversing different lengths of time intervals, which weakens local information and may result in decreased performance in weak signal regions. The change-point detection method's performance is easily disrupted by fluctuations in time series. Although the different-region balance method reduces the potential bias of sliding windows, its performance still depends on the choice of window sizes. Therefore, window sizes are crucial for sliding window methods, or rather, the appropriate detection regions are essential for accurate causal detection, yet it remains a challenge and has been rarely studied.

We address the issue of providing suitable detection regions for causal detection, and propose the Detection Windows based on Hidden Markov Model (HMMDW) method, which focuses on utilizing hidden Markov model (HMM) methods to provide appropriate detection regions. Our contribution includes designing the hidden state sequence of the hidden Markov model to correspond to suitable detection regions, designing that the fitting degrees of regression models are as the observation sequence of the hidden Markov model, and devising a method to conduct multiple causal tests and synthesize these results.

Section 2 introduces sliding window methods for exploring causal relationships, Sect. 3 designs a method for discovering time-varying causal relationships based on hidden Markov models, Sect. 4 presents experimental results and discussions, and finally, the conclusion.

2 Sliding Window for Causal Discovery Between Time Series

In order to detect the causal relationship between time series over time, the common sliding window method uses a window Wt of size U to move by the step along the time axis. At the same time, in the current window Wt, the Granger causality detection method is used to test whether there is a causal relationship between the time series in the window, and a series of the test results constitutes the time-varying causal relationship.

For time series $X = \{x_1, x_2, ..., x_T\}$ and $Y = \{y_1, y_2, ..., y_T\}$ with length T, an autoregressive model (called reduced model Mr) is used to describe the variation of time series Y within the current window Wt, and a regression model (called full model Mf) that includes time-lagged values of variables X and Y is used to predict the variation of time series Y. There is a Granger causality between the time series X and Y if Mf predicts the future values of Y more accurately than Mr does. The two regression models (with lag parameter, regression coefficients) fitting the time series X and Y are [8]:

$$\mathrm{Mf}: y_t = \sum_{l=1}^{L} a_l \cdot y_{t-l} + \sum_{l=1}^{L} b_l \cdot x_{t-l}$$

$$\mathrm{Mr}: y_t = \sum_{l=1}^{L} a_l \cdot y_{t-l} \quad . \tag{1}$$

The F-test is used to determine whether Mf is a more accurate predictor than Mr. For using F-test, the F statistic is constructed:

$$F = \frac{\left(SSE_{Mr} - SSE_{Mf}\right)/L}{SSE_{Mf}/(T - 2L - 1)}, \tag{2}$$

where the sum of squared prediction errors is $SSE = \sum_{t=1}^{T} (y_t - \hat{y}_t)^2$.

Under the null hypothesis (the prediction of Mf is not significantly better than Mr), if the F-value calculated from the time-series data is greater than the threshold at the level of significance α, the null hypothesis is rejected, and X is the cause of Y ($X \rightarrow Y$). Otherwise, it is judged that there is no causal relationship of $X \rightarrow Y$. Similarly one can judge whether Y is the cause of X.

3 HMM Detection Windows for Causal Discovery Between Time Series

In this section the HMMDW method is designed to detect time-varying causal relationships. Its detection framework is designed as follows: the time series of length T is partitioned into a number of disjoint regions, the kth region corresponds to the kth window W_k of size U_k; the Granger causality test is conducted within each window, resulting in a series of Granger test results that constitutes varying causal relationships over time. The task of partitioning the time series into a number of disjoint regions will be accomplished by the HMM approach.

3.1 Partitioning Time Series by Hidden Markov Model

Hidden Markov model has a hidden state variable, state transfer probabilities and an observation variable, where the hidden states are internal states that cannot be directly observed. It is assumed that the state at the current moment is only related to the state at the previous moment, the hidden states are transferred to each other according to the probability, and the observation sequence is generated by the hidden states with a certain probability.

Let the hidden variable Z have two states (the state value is 1 or 0), the $a_{ij} = P(z_{t,j}|z_{t-1,i})$ is the state transfer probability from state i at moment t-1 to state j at moment t. Let the observed value o_t of the observation variable O at moment t can be estimated by the observation probability distribution $b_t = P(o_t|z_t)$, and assume that the observation probability density obeys a normal distribution. Given the observation sequence, set the initial parameters of the HMM, the hidden state sequence in the HMM can be derived using the Baum-Welch algorithm and the Viterbi algorithm.

For the task of segmenting the time series, the effectiveness of fitting the time series with the full model Mf is first evaluated. In detail, a sliding window method is used to traverse the time series of length T, i.e., a window Wt of size U is used to move along the time axis at step size S; for each window Wt, the Mf is used to fit the time series in Wt, and the adjusted R-squares of the Mf fitting the time series is computed; thus, the sequence of adjusted R-squares is obtained.

Second, the sequence of adjusted R-squares is used as the observation sequence to solve the HMM, whose output is a sequence of hidden states (with value 1 or 0) of length T. The value 1 or 0 corresponds to high or low adjusted R-square.

Finally, every range of consecutive occurrences of 1 (or consecutive occurrences of 0) in the hidden state sequence is identified. A range with such continuous state values is called a consecutive-state region. A consecutive-state region corresponds to a segmentation region of the time series, so the time series is partitioned into a number of disjoint regions.

3.2 Causal Discovery with HMM Detection Windows

After the time series is partitioned into some disjoint regions, the subsequent work is to detect Granger causality in each partitioned region and record the detection results (with or without Granger causality), and the detection results of these regions form the time-varying causality. Here we do not distinct causal directions, any of the direction $X \rightarrow Y$ or $Y \rightarrow X$ is simply regarded as the existence of causality.

Since a segmented region may not be a complete causal region, and may contain non-causal regions, some bias may occur if the Granger causality test is performed once on all the data in a region. Here, we design a **regional causality test method** to perform multiple causality tests and synthesize these results for a partitioned region R, with the following steps:

1) Neighboring region merging: if the length V of segmented region R is smaller than threshold V_0, the region R expands by merging with its neighboring regions;

2) Test in region R: a Granger causality test is performed once on all the data in R, and the result (with causality or without causality at each time point) is recorded (noted as record A);

3) Region length checking: if the length $V/2 < V_0$, stop; otherwise, go to step 4);

4) Sliding window detection: a sliding window of size $V/2$ is used in region R to perform the Granger causality test, and record the results (noted as record B); then use a sliding window of size $V/3$ to perform Granger causality test and record the results (noted as record C);

5) Result synthesis: the majority-win voting method is used to decide whether there is a causal relationship or not at each time point on the region R, using the record A, B and C.

Finally, the main steps of the HMMDW method are summarized in Algorithm 1.

Algorithm 1: Detection Windows based on Hidden Markov Model (HMMDW)

Input: time series X and Y of length T, window size U, step size S, maximum time lag L.

1) Build regression models fitted to data: traverse the time series with a sliding window Wt of size U by step S, and in each window Wt, find the regression model Mf fitted to the data;

2) Evaluate the fitting effect of Mf: in each window Wt, calculate the coefficients of determination (adjusted R-squares) for the model Mf fitting the data, and obtain the adjusted R-square sequence of length T;

3) Solve the HMM: solve the HMM using the adjusted R-square sequence, and obtain the hidden state sequence of length T;

4) Segmentation of time series: find the consecutive-state regions, which correspond to segmented regions of the time series;

5) Causality testing of segmented regions: the regional causality test method is used to test every segmented region, and the test result (discriminant values corresponding to with or without causality) at each time point is recorded;

6) Fine-tuning of test results: perform Granger causality test with a sliding window (e.g., size $U_0=50$) for the provisional region (its length is $2U_0$) centered on the segmented region edge, and the test result is used to update discriminant values in the region (its length is U_0) centered on the region edge.

Output: T discriminant values (corresponding to with or without causality) at every time point

---

4 Experimental Result

This section provides performance experiments of the HMMDW method, compared with the conventional sliding window method (abbreviated as sliding window), the F-bound detection method, and the different-region balance method (abbreviated as region balance). The relevant parameters are set as follows: the default sliding step size is 5, the significance level of the F-test is 0.05, and the time lag parameter L gives the constraint that the length of either the sliding window or the detection region is not less than $3L + 2$. For the F-bound detection method, the time series is traversed by step size of 1, and the causal threshold is 0.9 times of the highest causality score.

The performance evaluation metric uses the accuracy ($100\% \times m/T$), which needs to count the number (m) of that predictions equal to the ground-truth at every time point, and then calculate the ratio of m to the length (T) of time series.

The simulated dataset was generated following reference [8]: stationary time series X and Y, their length is 500, time lag is 2, causal interval is $[s_1, s_2] = [150, 350]$. One dataset is generated for each experiment, including randomly generated noise, and the s_1 and s_2 are added with random integers from [-4,5]. For the sliding-window method and different-region balance method, due to the difficulty of choosing a suitable window size, one performance experiment includes three experiments with window sizes 20, 40, and 80 respectively, and gives the average of the three solutions. Table 1 shows the average results of 20 performance experiments.

Table 1. Accuracy of the methods to discover causal relationships on simulated datasets (%).

Standard deviation of noise for datasets	Sliding window	Region balance	F-bound detection	HMMDW
0.05	91.85	**92.62**	88.88	92.11
0.1	89.14	**92.54**	87.88	91.76
0.3	83.60	88.19	**92.30**	91.13
0.5	80.09	84.43	**89.72**	88.43

The real dataset Dropoff-tweet, from the Fig. 4 of reference [8], is the time series (with a sample size of 745) about the number of taxi drop-off passengers at the New York Convention Center and the number of audience comment posts over time in October 2012. When there is a comedy show (causal time interval [248, 337]) at the New York Convention Center, Dropoff sequence X of taxis is responsible for the change in the number of comment posts (Tweet sequence Y). The square root of data is computed as data preprocessing; the time lag $L = 5$. The task is to detect when there is causal relationship between the sequences, and the experimental results are listed in the Table 2.

The real dataset Tweet-pickup, from the Fig. 4 of reference [8], is a time series of the number of taxi pick-up passengers (sequence X) at the New York Convention Center in October 2012 versus the number of comment posts (Tweet sequence Y) over time. The Y causes the change in the X when there is a comedy show. The experimental results are presented in the Table 2.

Table 2. Accuracy of the methods to discover causal relationships on real datasets (%).

Datasets	Sliding window	Region balance	F-bound detection	HMMDW
Dropoff-tweet	90.38	91.54	94.09	**98.12**
Tweet-pickup	91.99	93.33	92.88	**95.44**

It can be seen from the tables above, the HMMDW method successfully reveals the causal relationships between time series over time. On the simulated dataset, the HMMDW method behaves in sub-optimal performance with stable performance; more importantly, the HMMDW method has a higher practical value than the different-region balance method, whose good performance need choose suitable but unknown window sizes. On real datasets with low noise, the HMMDW method outperforms the comparison methods. These results indicate that the HMMDW method outperforms the comparison methods in terms of comprehensive performance with high accuracy and stable performance, especially at lower noise levels.

In Table 3, the state regions from the Hidden Markov Model shows that the scope (from head to tail time points) of corresponding state region is close to a real causal region, which indicates that the HMMDW method provides suitable potential detection regions.

Table 3. State regions from HMM (head to tail time points of a region).

Datasets	Real causal regions	State regions from HMM	Total number of HMM regions
Dropoff-tweet	248 - 337	218 - 377	3
Tweet-pickup	248 - 337	223 - 347	8

It is known that the main weakness of the sliding window approach is that its performance is sensitive to window size: different sizes are likely to result in different performance, and inappropriate window sizes are likely to result in low performance. The proposed HMMDW method gains appropriate window sizes (detection regions) in advance, so that it can hold good performance of causality detection by using suitable detection regions.

5 Conclusion

In this paper, we propose the HMMDW method, which designs a Hidden Markov Modeling approach to partition the time series into a series of regions, and probes the causality in each region to obtain the changing causality. For the task of exploring changing causality between time series, our method provides reasonable detection regions and offers a solution to the difficult problem that the size of the detection regions is difficult to estimate in advance.

Acknowledgements. This work is supported in part by the Natural Science Foundation of Fujian Province of China under Grant No 2022J01656, and the National Natural Science Foundation of China under Grant No 62106049.

References

1. Ren, W., Han, M.: Survey on causality analysis of multivariate time series. Acta Automatica Sin. **47**(1), 64–78 (2021)
2. Granger, C.W.J.: Investigating causal relations by econometric models and cross-spectral methods. Econometrica **37**(3), 424–438 (1969)
3. Xing, S., Niu, J., Ren, T.: GCFormer: granger causality based attention mechanism for multivariate time series anomaly detection. IEEE International Conference on Data Mining, Shanghai, China, December 1–4, pp. 1433–1438 (2023)
4. Ren, W., Li, B., Han, M.: A novel granger causality method based on HSIC-Lasso for revealing nonlinear relationship between multivariate time series. Phys. A **541**, 123245 (2020)
5. Huang, B., Zhang, K., Gong, M., Glymour, C.: Causal discovery and forecasting in non-stationary environments with state-space models. In: The 36th International Conference on Machine Learning, Long Beach, California, USA, June 9–15, pp. 2901–2910 (2019)
6. Finkle, J.D., Wu, J.J., Bagheri, N.: Windowed Granger causal inference strategy improves discovery of gene regulatory networks. Proc. Natl. Acad. Sci. U.S.A. **115**(9), 2252–2257 (2018)
7. Chang, T., Tsai, S.L., Haga, K. A.: Uncovering the interrelationship between the U.S. stock and housing markets: a bootstrap rolling window Granger causality approach. Appl. Econ. **49**, 5841–5848 (2017)
8. Li, Z., Zheng, G., Agarwal, A., Xue, L., Lauvaux, T.: Discovery of causal time intervals. In: The Seventeenth SIAM International Conference on Data Mining. Westin Galleria Houston, Houston, Texas, USA. April 27–29, pp. 804–812 (2017)
9. Masnadi-shirazi, M., Maurya, M.R., Pao, G., Ke, E., Verma, I.M., Subramaniam, S.: Time varying causal network reconstruction of a mouse cell cycle. BMC Bioinform. **20**, 294 (2019)
10. Wang, K., Zeng, Y., Miao, Z.: Different-region balance method for exploring varying causal relations between time series. J. Electron. Inf. Technol. **43**(8), 2414–2420 (2021)

Real-World Implications of a Methodological Dilemma: Endogenous Confounding in Causal Decomposition Analysis

Ha-Joon Chung[1] and Guanglei Hong[2]([✉])

[1] Princeton University, Princeton, NJ 08544, USA
[2] The University of Chicago, Chicago, IL 60637, USA
ghong@uchicago.edu

Abstract. Should a set of hypothetical interventions eliminate the Black-White gap in malleable factors such as schooling, how much racial disparity in subsequent youth development would be reduced and how much would remain? To address this question, a causal decomposition analysis must credibly identify the causal impact of intervening on the malleable factors. However, major confounders such as parental SES are intermediate outcomes of systemic racism experienced over generations that place Black households and communities at a great disadvantage. For this reason, an intervention that attempts to eliminate the Black-White gap in schooling within levels of parental SES or other endogenous confounders has two major problems: (1) Such an attempt is infeasible when only one but not both racial groups are present within some extreme levels of parental SES. (2) Even if both groups are present within every level of parental SES, the attempt would nonetheless fail to eliminate the marginal Black-White gap in schooling due to the endogeneity of parental SES. We propose a novel solution that replaces the original scale of an endogenous confounder with a scale preserving one's within-group relative standing. A semiparametric weighting strategy then emulates an equity-oriented intervention. Our analysis of the NLSY 1997 data reveals real-world implications of the methodological dilemma posed by endogenous confounding.

Keywords: Causal inference · Disparity analysis · Mediation analysis

1 Introduction

Persistent and often drastic disparities in economic and social wellbeing between groups defined by gender, race/ethnicity, social class, caste, national origin, or urban vs. rural residential status have been of major concern to society and a focus of many domestic and international policies and programs promoting equality and equity. In applied statistics and econometrics, the conventional Blinder-Kitagawa-Oaxaca decomposition [1–3] descriptively decomposes a between-group mean difference in an outcome (denoted as Y) into a component associated with one or more explanatory factors (denoted as M that takes value m for $m \in \mathcal{M}$) and a second component that remains unexplained. However, the descriptive analysis does not answer the question of whether intervening on M would

X. Zhou and J. Jia (Eds.): PCIC 2024, CCIS 2200, pp. 49–64, 2025.
https://doi.org/10.1007/978-981-97-7812-6_5

subsequently reduce disparity in Y; thus, such an analysis contributes limited evidence to theoretical understanding or informing practical decision-making in choosing targets of potential interventions.

In contrast, an active area of research in the recent years is to emulate an experiment in which a hypothetical intervention would disrupt the mechanism that produces or reproduces disparity in the outcome between social groups. A *causal decomposition analysis* partitions the between-group disparity in Y into a component that could be reduced through conceivable or actual interventions on one or more malleable factors M and a component that would remain. The malleable factors M may not be a treatment itself but rather an immediate target of interventions. Elements of such interventions may or may not be well specified. Nonetheless, M represents a key process, possibly in a cumulative manner, through which these interventions may operate towards the goal of reducing disparity in the outcome. Investigations of the potential impacts of stylized hypothetical interventions allow researchers to garner insights on which social processes are responsible for the existing disparity and how much of the latter we may hope to reduce. The analysis aims to obtain a distribution of the counterfactual outcome for one or more groups in a counterfactual world in which stylized hypothetical interventions would have eliminated or at least greatly reduced the between-group gap in M in the population. The results are credible only when the causal relationship between M and Y within each group can be identified under plausible assumptions, typically by conditioning on the values of "baseline covariates." This is because when M is *as if* randomized among those of the same covariate values, for individuals who have not experienced a desired level m in the current world but would in the counterfactual world, their counterfactual outcome is to be inferred from a comparable group of individuals who have already experienced the desired level of m in the current world [4–15].

This study reveals the methodological dilemma posed by *endogenous confounding* in causal decomposition analysis, discusses its real-world implications, and proposes a solution. We highlight the distinction between two types of covariates that confound the $M \rightarrow Y$ relationship—exogeneous confounders vs. endogenous confounders. Exogeneous confounders (denoted by X) such as age or gender generally do not differ in distribution between the groups of interest. Hence the average *conditional* between-group difference in M is expected to be equal to the pre-existing *marginal* between-group difference in M. In contrast, endogenous confounding arises when a potential confounder (denoted by A) of the $M \rightarrow Y$ relationship distributes differently across the groups of interest. As the groups of concern are deemed to have unequal social status and differential access to resources and thereby displaying divergent trajectories in the past or even over many generations, endogenous confounders tend to be prevalent. Through mathematical derivation and graphical representation, we will show that the *conditional* between-group difference in M averaged over levels of A is no longer equal to the *marginal* between-group difference in M. As a result, a hypothetical intervention that eliminates the *conditional* between-group difference in M within levels of $A = a$ would fail to eliminate the *marginal* between-group difference in M. Hence, a causal decomposition analysis that intends to assess the potential impact of eliminating the *marginal* between-group difference in M would generally miss the mark when the analysis instead eliminates the *conditional* between-group difference in M. Furthermore, in one extreme case, the support for M may differ between the groups when conditioning on $A = a$; in another, a group of interest may not even exist within certain levels of A.

Hence the dilemma is that, if not conditioning on the values of an endogenous confounder A, the analyst would fail to identify the $M \rightarrow Y$ relationship within each group and thus unable to obtain the counterfactual outcome values needed for the causal decomposition analysis; yet if conditioning on the values of A, the analyst would obtain the counterfactual outcome values associated with a hypothetical world in which the *marginal* distribution of M would remain unequal between the groups. The issue is related to the well-known Simpson's Paradox [16, 17] and the so-called "collider bias" in the directed-acyclic-graphs (DAG) representation of causal inference [9, 18, 19]. Yet the discussion has been limited in the rapidly growing literature on causal decomposition analysis (see [13], for an exception).

This study clarifies the potential consequences for causal decomposition analysis when the analyst attempts to adjust for endogenous confounders in the same manner as adjustment for exogeneous confounders. To address this thorny problem, we propose a rank-preserving transformation of an endogenous confounder that removes the endogeneity; we then employ a semiparametric weighting approach to simulate a set of equity-oriented interventions. The paper clarifies key identification assumptions required by this approach and reveal important real-world implications. Analyzing the National Longitudinal Study of Youth 1997 (NLSY 97) data, we investigate the potential of reducing the racial gap in youth outcome through hypothetical interventions in earlier schooling. Naive interventions that overlook the racial disparity in parental, community, and school resources would fail to achieve racial equity in targeted malleable factors such as schooling. By adopting the rank-preserving strategy, we simulate a set of desired interventions that would instead effectively overcome race-based deprivations.

2 Background

2.1 Substantive Context

Despite race being a major source of variation in developmental experiences and outcomes, it is an ill-defined cause in the potential outcome framework [20, 21]. This is because as an 'immutable' attribute, race in general cannot be experimentally manipulated [22–24]. The literature has focused on the fact that racial inequalities arise not only from the attributes that are tied to systemic racism in the U.S. history but also from structural barriers and discriminatory practices in the current society that Blacks are subjected to [25, 26]. A meaningful causal question about race thus should attend to malleable factors that differentially affect Blacks versus Whites via social processes [27]. Such insights have motivated efforts in empirical investigations to decompose, for example, racial disparities in physical health or economic wellbeing into a portion that is mediated through certain socially constructed conditions (e.g., racial differences in schooling) and a portion that is not. Evidence from a credible decomposition analysis promises to advance social scientific theories and inform policies and intervention programs aimed at reducing disparities.

Prolonged disconnection from school and work represents major setbacks during the transition to adulthood and is a distinct feature of the developmental trajectories of many disadvantaged youths, especially those from a marginalized racial background [28, 29]. Differential schooling experiences are hypothesized mechanisms that widen the racial disparity (e.g., [30–33]. Yet youths from a marginalized background appear to be the ones who can reap the most benefit when engaged in education [34]. Hence equalizing educational opportunities and promoting educational equity have been advocated as major vehicles for propelling the upward mobility of racial minority youths. The empirical question is: To what extent could the racial disparity in youth disconnection from school and work be reduced through reducing the racial gap in earlier schooling?

Estimand. Let $S = 1$ denote Black youths and $S = 0$ for White youths. The racial disparity in youth disconnectedness from school and work in the current world is a descriptive estimand that does not involve counterfactual outcomes. It has the following definition: $\delta = E[Y(M)|S = 1] - E[Y(M)|S = 0]$. Let M represent the actual schooling experiences of each racial group in the current world while M^* represents the counterfactual earlier schooling experiences under hypothetical interventions. Corresponding to the substantive research questions, we define two estimands. The definition of these estimands is consistent with what is now consensus in the literature on causal decomposition analysis.

(1) Reduction in racial disparity in youth disconnectedness should the distribution of earlier schooling experiences of Black youths become comparable to that of Whites:
$$\delta_{reduced} = E[Y(M)|S = 1] - E[Y(M^*)|S = 1].$$
(2) Remaining racial disparity in the outcome despite the equalization efforts in schooling: $\delta_{remain} = E[Y(M^*)|S = 1] - E[Y(M)|S = 0]$. Here $E[Y(M^*)|S = 1]$ is the average counterfactual outcome in a hypothetical world in which the distribution of earlier school experiences among Blacks would become equal to that of Whites.

An intervention that changes the distribution of M among Blacks to resemble its distribution among Whites is very different from an intervention that changes the distribution of M for both Blacks and Whites. Henceforth we focus on the former type of intervention.

2.2 Methodological Challenges Due to Endogenous Confounding

As noted earlier, identification of the counterfactual quantity $E[Y(M^*)|S = 1]$ requires adjustment for two types of confounding covariates. In the current application, the first type (X) is independent of race (e.g., age and gender) while the second type (A) distributes differently across the racial groups (e.g., parental SES). Due to the legacy of racial discrimination in the United States, Black youths are disproportionately represented among those of the lower-SES. For example, for every dollar of net wealth of a typical white household, a typical Black household has only 6 cents [35]. On average, youths growing up in lower-SES households tend to struggle more in school and later become more disconnected from school and work. Figure 1 illustrates the conceptual relationships among S, M, Y, X, and A. Here only the relationship between M and Y is causal. Importantly, A is downstream from S while X is not.

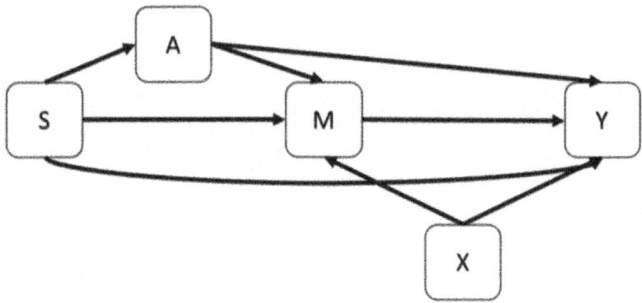

Fig. 1. The Default DAG

When the causal relationship between M and Y within each racial group is confounded by X only, we have that $pr(X = x|S = s) = pr(X = x)$ for $s = 0, 1$. In a hypothetical world in which the racial disparity in M were eliminated within levels of $X = x$, that is, $E[M^*|S = 1, X = x] = E[M|S = 0, X = x]$, the counterfactual distribution of M among Blacks would become equal to that of Whites. This is because $E[M^*|S = 1] - E[M|S = 0] = \sum_x E[M^*|S = 1, X = x]pr(X = x|S = 1) - \sum_x E[M|S = 0, X = x]pr(X = x|S = 0) = \sum_x \{E[M^*|S = 1, X = x] - E[M|S 0, X = x]\} pr(X = x) = 0$.

However, this result does not hold in the presence of endogenous confounders A given that $pr(A = a|S = s) \neq pr(A = a)$ for $s = 0, 1$. One may envision a hypothetical world in which the racial disparity in M were eliminated within levels of $A = a$, that is, $E[M^*|S = 1, A = a] = E[M|S = 0, A = a]$. We can show that the counterfactual distribution of M among Blacks would not become equal to that of Whites. This is because $E[M^*|S = 1] - E[M|S = 0] = \sum_a E[M^*|S = 1, A = a]pr(A = a|S = 1) - \sum_a E[M|S = 0, A = a]pr(A = a|S = 0) \neq \sum_a \{E[M^*|S = 1, A = a] - E[M|S = 0, A = a]\}pr(A = a)$. The discrepancy between the average conditional difference and the marginal difference between Blacks and Whites in the counterfactual distribution of M is $\sum_a E[M^*|S = 1, A = a] [pr(A = a|S = 1) - pr(A = a)] - \sum_a E[M|S = 0, A = a][pr(A = a|S = 0) - pr(A = a)]$.

To illustrate, let $A = 1$ if a student grew up in a higher-SES household and $A = 0$ if the student came from a lower-SES household. Higher-SES households and lower-SES households each constitute 50% of the combined population. However, this is not true within each racial group. Rather, an analysis of the NLSY97 data shows that only about 35% of the Black households are of higher-SES in contrast with about 70% of the White households being higher-SES. This pattern is shown in Fig. 2 where SES on its original scale is centered at mean zero while the within-race percentile values of SES range from 0 to 1.

Moreover, the support for an endogenous confounder such as SES may differ between the racial groups. Thompson and Suarez [39] showed that the White, Black, and Hispanic distributions of wealth do not completely overlap and cautioned that the Kitagawa-Blinder-Oaxaca method would extrapolate beyond the observed wealth range for the minority group simply because there are no minority individuals who could be matched to the richest majority individuals. Indeed, in the United States, Blacks historically have

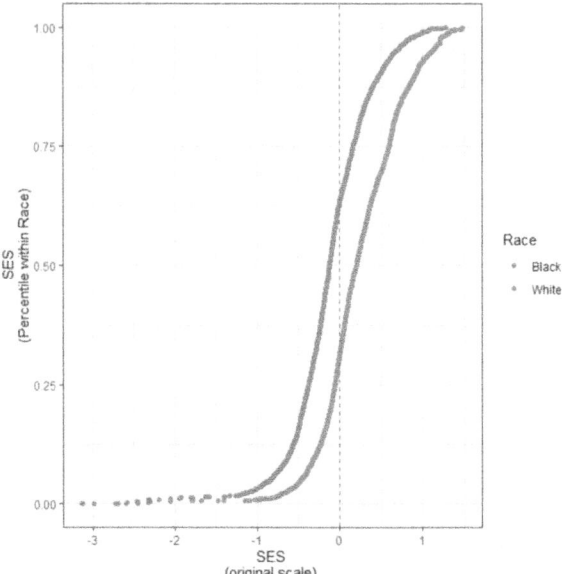

Fig. 2. Racial Gap in SES Distribution

had lower rates and magnitude of intergenerational transfers [40] due to a long history of racialized polices that segregated Blacks in low-income communities and restricted their access to assets [41]. The typical White family owns eight to ten times the wealth of the typical Black family and five times the wealth of the typical Hispanic family [42, 43]. Even more shockingly, the four hundred richest American billionaires who are almost exclusively White have more total wealth than all 10 million Black households combined [44].

2.3 Existing Strategies

Past research has suggested several different intervention frameworks for a causal decomposition analysis. They are characterized as anti-discriminatory interventions, lottery-type interventions, and affirmative action (AA)-type interventions.

Anti-Discriminatory Intervention. Due to racial prejudice and profiling, Black students are disproportionately labeled as "learning disabled," placed in non-academic tracks, and subjected to disciplinary measures, all of which are known to increase the risk of high school dropout. Let $M = 1$ if a student has completed 12 or more years of schooling and 0 otherwise. To counteract a racially biased system that creates differential schooling experiences between Black and White students from similar backgrounds measured by exogenous covariates X and endogenous covariates A, the objective of an anti-discriminatory intervention is to achieve $\Pr[M^* = 1|S = 0, A = a, X = x] = \Pr[M^* = 1|S = 1, A = a, X = x]$, so that the distribution of M^* would be the same for those with the same covariate values a and x regardless of race. However, as we have reasoned earlier, equating the conditional distribution or the conditional odds of M

between Blacks and Whites within levels of $A = a$ would fail to achieve racial equity in M. Based on the information about high school completion rate in the NLSY97 data, Table 1 shows that even though the counterfactual high school completion rate among Blacks is made equal to the actual high school completion rate among Whites within each level of parental SES, the counterfactual high school completion rate among Blacks when averaged over the SES levels is only 68%, considerably lower than the marginal high school completion rate among Whites, which is 77%. This is due to the under-representation of Blacks among high-SES families and their over-representation among low-SES families. The anti-discriminatory intervention would fail to remove the racial gap in M^* transmitted indirectly through parental SES. A similar pattern holds when conditioning on exogenous confounders such as age and gender.

Table 1. Actual vs. Counterfactual Racial Gap in High School Completion Rate

		Blacks	Whites	Racial Gap
Higher-SES ($A = 1$)	Actual	0.78	0.84	−0.06
	Counterfactual	0.84	0.84	0
Lower-SES ($A = 0$)	Actual	0.56	0.60	−0.04
	Counterfactual	0.60	0.60	0
Combined	Actual	0.64	0.77	−0.13
	Counterfactual	0.68	0.77	−0.09

Note:$E[M^*|S = 1] = E[M^*|S = 1, A = 1]pr(A = 1|S = 1) + E[M^*|S = 1, A = 0]pr(A = 0|S = 1) = 0.84 \times 0.35 + 0.60 \times 0.65 \approx 0.68$.

Lottery-Type Intervention. Jackson and colleagues [6, 7] proposed a stochastic inter-vention for Blacks that takes a random draw from Whites' distribution of M either unconditionally or conditionally among those who share the same values of $X = x$ and $A = a$ with the goal of equating the distribution between groups to achieve either $\Pr[M^* = 1|S] = \Pr[M = 1]$ or $\Pr[M^* = 1|S, X, A] = \Pr[M = 1|X]$. As a result, M^* becomes independent of S, A, and X. Opacic and colleagues [36] described this type of stochastic intervention as "lottery-type equalization." In the example shown earlier in Table 1 a lottery-type intervention would simulate a counterfactual world in which higher-SES and lower-SES Blacks would have the same high school completion rate 0.77. However, random assignments that distort the mechanisms through which values of M are allocated as a function of A within a racial group is likely unrealistic and thus infeasible. A hypothetical intervention that attempts to equate all Black students' school-ing experiences with that of a typical White student may even ironically exacerbate rather than reduce the racial gap among students from high-SES families.

Affirmative Action (AA)-Type Intervention. Others have proposed an AA-type inter-vention that equalizes the conditional odds of a binary M between race [36, 37]. Adopting the dynamic intervention method introduced by Kennedy [38], they alter the distribution of M by adding a group-specific constant δ_s to the log-odds of M without changing the way it depends on A and X:

$$\log \frac{\Pr[M^* = 1|s, a, x]}{\Pr[M^* = 0|s, a, x]} = \delta_s + \log \frac{\Pr[M = 1|s, a, x]}{\Pr[M = 0|s, a, x]},$$

where δ_s for $s = 0, 1$ is to be numerically solved such that $\Pr[M^* = 1|S = 0] = \Pr[M^* = 1|S = 1] = \Pr[M = 1]$. These researchers argued that an AA-type intervention is designed to change the distribution of M uniformly within a racial group. Yet we can show that the impact of such an intervention on M will differ across levels of A. Given the functional form of log-odds being sigmoid, the added intercept δ_s will translate to a greater impact for those whose probability of $M = 1$ is closer to 0.5 prior to the intervention.

3 Methodological Approach

3.1 Equity-Oriented Intervention via Rank-Preserving Transformation of Endogenous Confounders

Focusing on schooling experiences as malleable factors M, we envision equity-oriented interventions designed to achieve racial equity in schooling in the counterfactual world. Such interventions must overcome racial disadvantages in M associated with the endogenous confounders. Our proposed alternative strategy involves a rank-preserving transformation of the endogenous confounder A. Let A_s denote the standardized score or the percentile score of A representing an individual's relative standing within racial group s for $s = 0, 1$. Most importantly, A_s is not endogenous to S; thus adjusting for A_s along with X would not evoke Simpson's paradox. The issue of inadequate common support with regards to A is also resolved. Unlike most other types of interventions described above, this intervention resembles an equity-oriented intervention aiming to overcome the standing racial (dis)advantage in A as we make $Pr(M^* = m|S = 1, X = x, A_{s=1} = a) = Pr(M = m|S = 0, X = x, A_{s=0} = a)$ while preserving the within-group predictive relationship between X and M and between A and M.

However, it is conceivable that certain levels of schooling are available to Whites but denied to their Black counterparts who share similar relative standing in SES within their respective racial groups. When this is the case, the racial disparity in schooling cannot be completely eliminated by the hypothetical intervention simply because, in the current world, there is no counterfactual information for Blacks associated with the levels of schooling that have been preserved as a privilege for Whites. When there is incomplete common support for M between the racial groups, especially when certain values of $M = m$ are observed among Whites but not Blacks who share the same covariate values x and a, the counterfactual $E[Y(M^* = m)|S = 1, X = x, A_{s=1} = a]$ cannot be identified for the Blacks. This is the case when an equity-oriented intervention would reduce but not eliminate the racial gap in M.

3.2 Identification Strategy

Our identification strategy is to simulate for Blacks their distributions of M^* and $Y(M^*)$ in the counterfactual world through semiparametric weighting. We simulate a hypothetical world in which the counterfactual distribution of M^* for Black students would be made equal to the observed distribution of M among their White counterparts who share the same covariate values $X = x$ and $A_s = a$. The weight is a ratio of Blacks' conditional distribution of M^* in the counterfactual world to their conditional distribution of M in the current world conditioning on X and A_s. For Blacks with $M = m$, $X = x$, and $A_{s=1} = a$,

$$W = \frac{Pr(M^* = m | S = 1, X = x, A_{s=1} = a)}{Pr(M = m | S = 1, X = x, A_{s=1} = a)}.$$

We envision a counterfactual world in which, for Blacks with covariate values $X = x$ and $A_{s=1} = a$, a set of hypothetical interventions would transform the distribution of their M^* to be the same as the distribution of M of their White counterparts who share covariate values $X = x$ and $A_{s=0} = a$, the above weight is equal to

$$W = \frac{Pr(M = m | S = 0, X = x, A_{s=0} = a)}{Pr(M = m | S = 1, X = x, A_{s=1} = a)} = \frac{[1 - pr(S = 1 | M = m, X = x, A_{s=0} = a)]pr(S = 1 | X = x, A_{s=1} = a)}{pr(S = 1 | M = m, X = x, A_{s=1} = a)[1 - pr(S = 1 | X = x, A_{s=0} = a)]}. \tag{1}$$

We derive the last equation by applying Bayes Theorem. This form of the weight easily accommodates multidimensional M.

Theorem. The conditional average counterfactual outcome $E[Y(M^* = m) | S = 1, X = x, A_{s=1} = a]$ for Blacks is identified by the conditional average observed outcome of their Black counterparts $E[Y(M = m) | S = 1, X = x, A_{s=1} = a]$ under the assumptions of conditional exchangeability and positivity:

$$E\big[Y(M^*) | S = 1\big] = E[WY | S = 1].$$

Assumption 1. Conditional Exchangeability.

$$f\big(Y(M^* = m) = y | S = s, M^* = m, X = x, A_s = a\big) = f(Y(M = m) = y | S = s, M = m, X = x, A_s = a)$$

In racial group s, the causal relationship between M and Y is no longer confounded within levels of x and a. Under this assumption, Blacks who would display $M^* = m$ in the counterfactual world are exchangeable to and are thus expected to have the same distribution of Y as Blacks who share the same covariate values x and a and already display $M = m$ in the current world. Assumption 1 is violated if there exist unmeasured confounders of the M-Y relationship within a racial group, in which case the results of the causal decomposition analysis will be biased.

Assumption 2. Positivity.

$$pr(M = m | S = s, X = x, A_s = a) > 0$$

After the rank-preserving transformation of endogenous confounders, within levels of x and a, we assume that individuals who display $M = m$ in racial group $S = 1$ can find counterparts in racial group $S = 0$ who also display $M = m$; the same is true vice versa. This assumption is violated, for example, if the support for M differs between race in a subpopulation defined by x and a. When Assumption 2 is violated, the counterfactual information $Y(M^*)$ would be lacking for Blacks if none of them in the current world have experienced the same level of schooling as some of their White counterparts.

Yu and Elwert [45] pointed out that the existing methods for causal decomposition analysis tend to conflate the dual function a hypothetical intervention plays, namely, (1) the reassignment of M within each group, and (2) the equalization of the distribution of M between groups. Putting it differently, for an intervention to introduce a counterfactual distribution of M, it often needs to specify a new selection mechanism that leverages possible heterogeneity of the effect of M on Y. They criticize previous research for placing an exclusive focus on equalizing the distribution of M between groups without contemplating the changes implied for the selection mechanism. Clarifying the nature of the new selection regime is indeed crucial for understanding the substantive and policy-relevant implications of a proposed intervention. Under Assumptions 1 and 2, covariates X and A include all those that modify the M-Y relationship. Hence, we consider within-group reassignment of M that changes its distribution within levels of these covariates.

4   Sample and Measurement

The National Longitudinal Survey of Youth 1997 (NLSY 97) is comprised of a nationally representative sample of 8,984 youth born between 1980 and 1984. The survey had 19 waves including the baseline in 1997, annual follow-ups until 2011, and biennial follow-ups thereafter. This study uses the baseline and follow-ups until 2015. For simplicity, we truncated the data to include only the youths identified as non-Hispanic black and whites which drives down the sample size to 7,000. There are about 1,400 individuals in each of the five consecutive birth cohorts.

Our outcome measure is maximum duration of disconnection from school and work in 2007. It captures the amount of time spent not being engaged in education and training nor being active in the labor force (i.e., neither employed nor searching for jobs). We measure this by using monthly or weekly records of youth's education and employment history to calculate the number of consecutive weeks spent disengaged in a given year.

Measures of schooling experiences between 1998 and 2006 include 8th grade GPA, high school GPA, # days suspended from school each year, # months missed school each year, # grades repeated or skipped, and the highest level of education attained.

We conduct a principal component analysis weighted by survey weights and thereby creating a continuous composite measure of parent socioeconomic status that incorporates information on household income, household wealth, and parents' education at the baseline. We then standardize this endogenous covariate within each racial group.

A considerable portion of the NLSY97 sample are missing due to attrition, non-responses, or data entry error. Depending on waves, key variables have 22%–48% of the values missing. For cognitive tests administered to a small subsample, up to 85%

of the sample is missing. To account for systemic missingness and to preserve the statistical power, we used multiple imputation to impute the missing values. The repeated
measurements across consecutive survey years increase the plausibility of the missing at
random (MAR) assumption [46] that is necessary for the imputation. We use the MICE
(v.3.14.0) package in R to generate five independently imputed datasets. After running
the analyses on these datasets, we apply Rubin's rule [47] to pool the estimates.

According to Table 2, Blacks and Whites display similar distributions in the exogeneous covariates and different distributions in the endogenous covariates. Table 3 compares between Blacks and Whites the distribution of each intervening factor M as well as
that of the outcome Y. Blacks lag behind Whites on average in educational achievement
and attainment. Relative to Whites, Blacks tend to be underserved by schools given that
Black students were much less likely to enroll in the academic track in high school and
were instead suspended for twice as many days as Whites. By the time that they reached
early adulthood in Year 2007, the average maximum duration of disconnection from
school and work for Blacks was about three weeks more than that for Whites.

Table 2. Exogeneous vs. Endogenous Covariates

	Race	
	Blacks	Whites
	Mean (SD)	
Exogeneous Covariates (X)		
Male (%)	49	49
Citizenship (%)	95.3	97.3
Endogenous Confounders (A)		
Parent's Income (logged)	9.78 (1.69)	10.7 (1.18)
Parent's Wealth (logged)	11.8 (0.57)	13.9 (0.29)
Parent's Years of Education	12.8 (2.26)	14.2 (2.62)
Mother's Age at 1st Birth	21.3 (5.18)	23.8 (4.67)

5   Preliminary Analytic Results

We analyze logistic regression models predicting S as a function of X and A_s with or
without controlling for M. The conditional probabilities are elements for computing the
weight defined in Eq. (1) for Black youths; in contrast, the weight for White youths is
simply equal to 1. The weight for the causal decomposition analysis is then multiplied
by the survey weight.

Table 3. Racial Gaps in the Intervening Factors and the Outcome

	Race	
	Black	White
	Mean (SD)	
Intervening Factors (M)		
Academic Track (%)	16	26
8th Grade GPA	2.62 (0.79)	2.94 (0.85)
HS GPA	2.58 (0.79)	2.93 (0.82)
Years of Education by age 20	11.6 (1.86)	12.1 (1.76)
Days Suspended from School by age 20	9.53 (26.6)	4.73 (19.1)
Outcome (Y)		
Maximum Duration of Disconnection (weeks)	7.66 (15.6)	4.51 (11.8)

In the presence of endogenous confounding, our proposed rank-preserving strategy enables the simulation of equity-oriented interventions that minimize the racial gap in M. In contrast, a simulation of anti-discrimination interventions that overlook the key distinction between endogenous and exogenous confounders would fail to achieve racial equity in M. We illustrate with two examples of M. Figure 3a shows the actual racial gap in academic track participation rate in high school under no intervention, the counterfactual racial gap under an anti-discrimination intervention, and that under an equity-oriented intervention. Figure 3b shows the corresponding actual or counterfactual racial gaps in high school GPA.

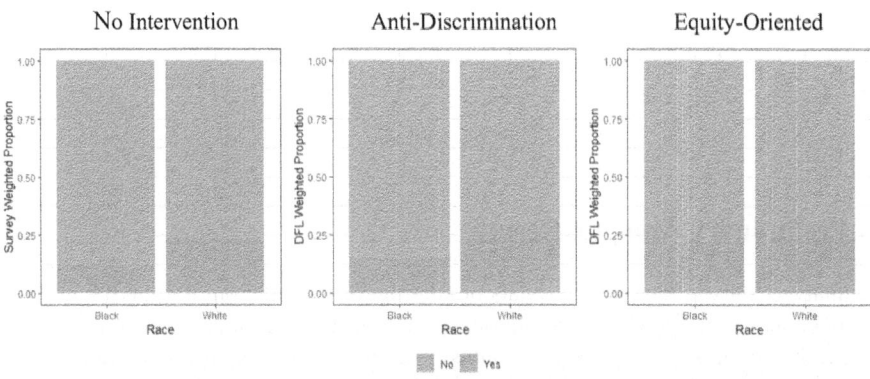

Fig. 3a. Actual and Counterfactual Racial Gaps in Academic Track Participation Rate

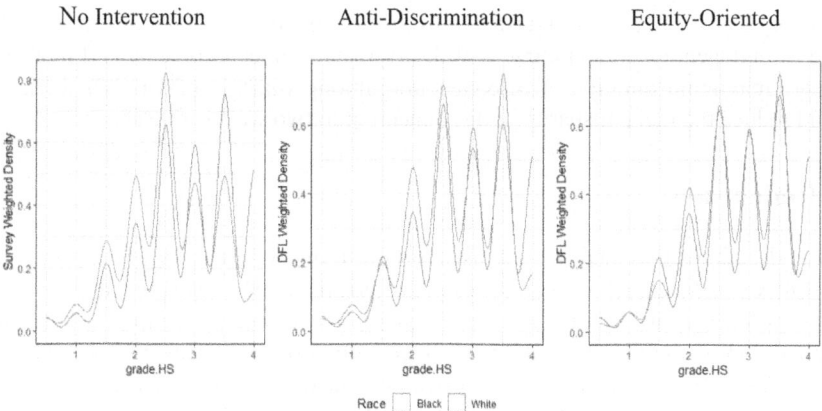

Fig. 3b. Actual and Counterfactual Racial Gaps in High School GPA

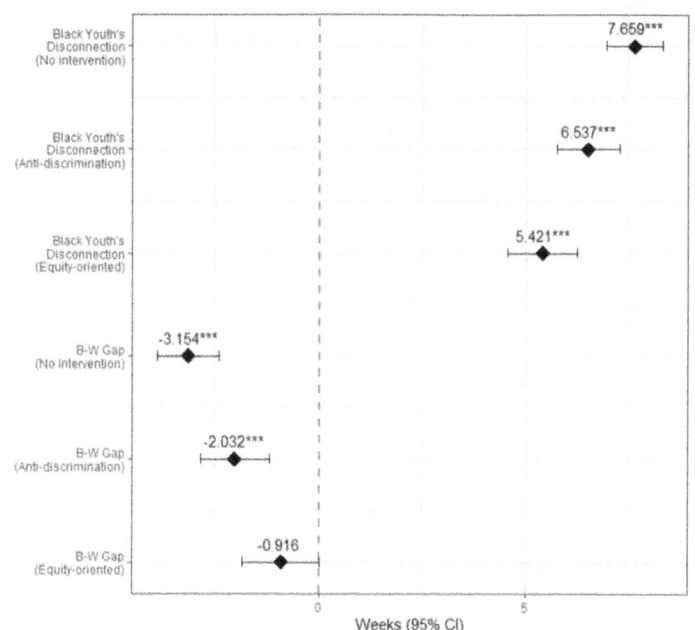

Fig. 4. Reduction in the Racial Disparity in Youth Disconnection under Equity-Oriented vs. Anti-Discrimination Interventions

The effect of equity-oriented interventions targeting schoolings experiences on youth disconnection is shown in Fig. 4. In a counterfactual world, if Black youths would have the same schooling experiences as their White counterparts who share the same covariate values after the rank-preserving transformation, their average maximum duration of disconnection would be reduced from 7.7 weeks to 5.4 weeks; the racial disparity in disconnection would be reduced from three weeks to only about one week, the latter becoming statistically indistinguishable from zero. In contrast, without addressing the issue of endogenous confounding, anti-discrimination interventions would reduce the average maximum duration of disconnection among Black youths to 6.5 weeks and would reduce the racial disparity in disconnection to two weeks instead.

6   Conclusion

Causal decomposition analysis promises to inform interventions for promoting equity. The validity and theoretical significance of analytic results are contingent upon the plausibility of identification assumptions. Yet even in the absence of omitted confounding, the analyst must appropriately handle endogenous confounders. This study builds on and contributes to the latest literature in this field. A causal decomposition analysis that simulates anti-discrimination interventions typically chooses to eliminate the conditional racial differences in the targeted malleable factors within levels of endogenous as well as exogenous confounders. We clarify that such interventions would fail to achieve racial equity in these factors. We innovatively address the issue of endogenous confounding by replacing the original scale with a scale representing one's within-group relative standing in the covariate, which enables the analyst to simulate an equity-oriented intervention.

In future analysis, we will incorporate this framework into a multi-cohort comparison strategy [28] in assessing the impact of the Great Recession on youth disconnection. The current application has conflated birth cohorts and has not accounted for age differences in the measures of schooling experiences. Such considerations will be undertaken in future analysis that focuses on how the hypothetical interventions in education may provide protection to Black youths and potentially alleviate the adversary impact of the Great Recession for them.

References

1. Blinder, A.S.: Wage discrimination: reduced form and structural estimates. J. Hum. Resour. **8**(4), 436–455 (1973). https://doi.org/10.2307/144855
2. Kitagawa, E.M.: Components of a difference between two rates. J. Am. Stat. Assoc. **50**(272), 1168–1194 (1955). https://doi.org/10.2307/2281213
3. Oaxaca, R.: Male-female wage differentials in urban labor markets. Int. Econ. Rev. **14**(3), 693–709 (1973). https://doi.org/10.2307/2525981
4. DiNardo, J., Fortin, N.M., Lemieux, T.: Labor market institutions and the distribution of wages, 1973–1992: a semiparametric approach. Econometrica **64**(5), 1001–1044 (1996). https://doi.org/10.2307/2171954
5. Huber, M.: Causal pitfalls in the decomposition of wage gaps. J. Bus. Econ. Statist. **33**(2), 179–191 (2015). https://doi.org/10.1080/07350015.2014.937437
6. Jackson, J.W.: Meaningful causal decompositions in health equity research: definition, identification, and estimation through a weighting framework. Epidemiology **32**(2), 282–290 (2021). https://doi.org/10.1097/EDE.0000000000001319
7. Jackson, J.W., VanderWeele, T.J.: Decomposition analysis to identify intervention targets for reducing disparities. Epidemiology **29**(6), 825–835 (2018). https://doi.org/10.1097/EDE.000 0000000000901
8. Li, F., Li, F.: Using propensity scores for racial disparities analysis. Observ. Stud. **9**(1), 59–68 (2023)

9. Lundberg, I.: The gap-closing estimand: a causal approach to study interventions that close disparities across social categories. Sociol. Meth. Res. 004912412110557 (2022). https://doi.org/10.1177/00491241211055769
10. Park, S., Kang, S., Lee, C.: Choosing an optimal method for causal decomposition analysis with continuous outcomes: a review and simulation study. Sociol. Methodol. **54**(1), 92–117 (2023). https://doi.org/10.1177/00811750231183711
11. Park, S., Kang, S., Lee, C., Ma, S.: Sensitivity analysis for causal decomposition analysis: assessing robustness toward omitted variable bias. J. Causal Inferen. **11**(1), 20220031 (2023). https://doi.org/10.1515/jci-2022-0031
12. VanderWeele, T.J., Robinson, W.R.: On the causal interpretation of race in regressions adjusting for confounding and mediating variables. Epidemiology **25**(4), 473–484 (2014). https://doi.org/10.1097/EDE.0000000000000105
13. Yamaguchi, K.: Decomposition of gender or racial inequality with endogenous intervening covariates: an extension of the DiNardo-fortin-lemieux method. Sociol. Methodol. **45**(1), 388–428 (2015). https://doi.org/10.1177/0081175015583985
14. Yamaguchi, K.: Decomposition analysis of segregation. Sociol. Methodol. **47**(1), 246–273 (2017). https://doi.org/10.1177/0081175017692625
15. Yamaguchi, K.: Determinants of the gender gap in the proportion of managers among white-collar regular employees. In: Yamaguchi, K. (ed.) Gender Inequalities in the Japanese Workplace and Employment: Theories and Empirical Evidence, pp. 47–81. Springer Singapore, Singapore (2019). https://doi.org/10.1007/978-981-13-7681-8_2
16. Lindley, D.V., Novick, M.R.: The role of exchangeability in inference. Ann. Statist. **9**(1) (1981). https://doi.org/10.1214/aos/1176345331
17. Pearl, J.: Comment: understanding simpson's paradox. Am. Stat. **68**(1), 8–13 (2014)
18. Elwert, F., Winship, C.: Endogenous selection bias: the problem of conditioning on a collider variable. Ann. Rev. Sociol. **40**, 31–53 (2014)
19. Greenland, S., Pearl, J., Robins, J.M.: Causal diagrams for epidemiologic research. Epidemiology **10**(1), 37–48 (1999)
20. Neyman, J.: On the Application of Probability Theory to Agricultural Experiments. Essay on Principles. Section 9. (Translated and Edited by D. M. Dabrowska and T. P. Speed from the Polish Original, Which Appeared in Roczniki Nauk Rolniczych Tom X). Ann. Agric. Sci. **10**, 1–51 (1923). (The translation was published in 1990 in Statistical Science 5(4): 465–480)
21. Rubin, D.B.: Multiple imputations in sample surveys" a phenomenological bayesian approach to nonresponse. In: Proceedings of the Survey Research Methods Section, pp. 20–34. American Statistical Association, Washington, DC (1978)
22. Rubin, D.B.: Comment: which ifs have causal answers. J. Am. Statist. Assoc. **81**, 396, 961–962 (1986). https://doi.org/10.1080/01621459.1986.10478355
23. Holland, P.: Statistics and causal inference. J. Am. Stat. Assoc. **81**(396), 945–960 (1986). https://doi.org/10.1080/01621459.1986.10478354
24. Holland, P.: Causation and race. ETS Res. Report Ser. **2003**(1), i–21 (2003). https://doi.org/10.1002/j.2333-8504.2003.tb01895.x
25. Destin, M.: A path to advance research on identity and socioeconomic opportunity. Am. Psychol. **74**, 1071–1079 (2019). https://doi.org/10.1037/amp0000514
26. Tilly, C.: Durable inequality. University of California Press, Berkeley, Calif (1998)
27. Sen, M., Wasow, O.: Race as a bundle of sticks: designs that estimate effects of seemingly immutable characteristics. Annu. Rev. Polit. Sci. **19**(1), 499–522 (2016). https://doi.org/10.1146/annurev-polisci-032015-010015
28. Hong, G., Chung, H.-J.: Assessing the impact of the great recession on the transition to adulthood. Sociol. Meth. Res. 004912412211138 (2022). https://doi.org/10.1177/00491241221113871

29. Shanahan, M.J.: Pathways to adulthood in changing societies: variability and mechanisms in life course perspective. Ann. Rev. Sociol. **26**(1), 667–692 (2000). https://doi.org/10.1146/ann urev.soc.26.1.667

30. Conwell, J.A.: Diverging disparities: race, parental income, and children's math scores, 1960 to 2009. Sociol. Educ. **94**(2), 124–142 (2021). https://doi.org/10.1177/0038040720963279

31. Hanushek, E.A., Peterson, P.E., Talpey, L.M., Woessmann, L.: The achievement gap fails to close. Educ. Next **19**(3), 8–17 (2019)

32. Jarvis, S.N., Okonofua, J.A.: School deferred: when bias affects school leaders. Soc. Psychol. Personal. Sci. **11**(4), 492–498 (2020). https://doi.org/10.1177/1948550619875150

33. Jencks, C., Phillips, M.: The black-white test scope gap: why it persists and what can be done. Brook. Rev. **16**(2), 24–27 (1998). https://doi.org/10.2307/20080778

34. Brand, J.E., Xie, Y.: Who benefits most from college? evidence for negative selection in heterogeneous economic returns to higher education. Am. Sociol. Rev. **75**(2), 273–302 (2010). https://doi.org/10.2307/27801525

35. US Census Bureau: Wealth and Asset Ownership for Households, by Type of Asset and Selected Characteristics: 2014 (2014). https://www.census.gov/data/tables/2014/demo/wea lth/wealth-asset-ownership.html

36. Opacic, A., Wei, L., Zhou, X.: Disparity analysis: a tale of two approaches. Working Paper (2023)

37. Zhou, X., Pan, G.: Higher education and the black-white earnings gap. Am. Sociol. Rev. **88**(1), 154–188 (2023). https://doi.org/10.1177/00031224221141887

38. Kennedy, E.H.: Nonparametric causal effects based on incremental propensity score interventions. J. Am. Stat. Assoc. **114**(526), 645–656 (2019). https://doi.org/10.1080/01621459. 2017.1422737

39. Thompson, J.P., Suarez, G.A.: Updating the racial wealth gap. Finan. Econ. Discuss. Ser. **2015**(076r1) (2017). https://doi.org/10.17016/feds.2015.076r1

40. Spilerman, S.: Wealth and stratification processes. Ann. Rev. Sociol. **26**(1), 497–524 (2000)

41. Oliver, M.L., Shapiro, T.M.: Black wealth/white wealth. In: Grusky, D., Grusky, D.B., Szelényi, S. (eds.) The Inequality Reader: Contemporary and Foundational Readings in Race, Class, and Gender, pp. 296–303. Routledge (2018). https://doi.org/10.4324/978042949446 8-33

42. Bhutta, N., Chang, A.C., Dettling, L.J., Hsu, J.W.: Disparities in wealth by race and ethnicity in the 2019 survey of consumer finances. FEDS Notes **2020**(2797) (2020). https://doi.org/10. 17016/2380-7172.2797

43. McIntosh, K., Moss, E., Nunn, R., Shambaugh, J.: Examining the black-white wealth gap. Brookings (blog). February 27, 2020. https://www.brookings.edu/blog/up-front/2020/02/27/ examining-the-black-white-wealth-gap/

44. Williamson, V.: Closing the Racial Wealth Gap Requires Heavy, Progressive Taxation of Wealth. Brookings (blog). December 9, 2020. https://www.brookings.edu/research/closing-the-racial-wealth-gap-requires-heavy-progressive-taxation-of-wealth/

45. Yu, A., Elwert, F.: Nonparametric causal decomposition of group disparities. Working paper (2024)

46. Little, R.J.A., Rubin, D.B.: Statistical Analysis with Missing Data. 2nd edn. Wiley Series in Probability and Statistics. Wiley, New York

47. Rubin, D.B.: Multiple Imputation for Nonresponse in Surveys. Wiley Series in Probability and Mathematical Statistics. Applied Probability and Statistics, pp. 0271–6232. Wiley, New York

Evaluation Criteria for Causal Discovery Without Ground-Truth Graphs

Lei Wang, Shanshan Huang, Liao Jun, and Li Liu$^{(\boxtimes)}$

School of Big Data and Software Engineering, Chongqing University,
Chongqing 401331, China
20222401006@stu.cqu.edu.cn, {shanshanhuang,liaojun,dcsliuli}@cqu.edu.cn

Abstract. Causal discovery in causal learning aims to derive causal graphs from observational data. However, collecting natural data is challenging and costly, so prior research has predominantly relied on synthetic datasets for validation. These synthetic or semi-real datasets, controlled artificially, may not fully reflect an algorithm's performance in real-world scenarios. Therefore, we proposed a method for evaluating causal discovery in the absence of a ground truth causal graph. First, we divided the data into training and test sets, then performed causal discovery on the training set to obtain causal graphs. We subsequently conducted Markov blanket tests on the causal graphs using the test set and determined the causal direction of each edge using multiple methods. Finally, we integrated the results of these methods through weighted voting to achieve the final accuracy. Experiments on both synthetic and real datasets demonstrated that our proposed method can reflect true accuracy to some extent.

Keywords: Causal Discovery · Markov Blanket · Causal Graphical Model · Condition Independence Testing

1 Introduction

Deep learning has achieved remarkable breakthroughs in various fields such as computer vision [16], natural language processing [4], and healthcare [7], driving rapid advancements in artificial intelligence. However, despite their ability to recognize complex patterns in large datasets, these models often raise concerns regarding the understanding of causal relationships because they primarily capture correlations. This is particularly problematic in healthcare, where misinterpreting correlation as causation can lead to misleading medical interventions. To address these issues, causal learning has become a focal point. Unlike deep learning, which focuses solely on correlations, causal learning aims to uncover the causal structures among variables, addressing issues related to interventions and counterfactual reasoning. A core branch of causal learning, causal discovery, seeks to map out the causal relationships among variables using Directed Acyclic Graphs (DAGs) [11], supporting more accurate predictions and decision-making.

X.-H. Zhou and J. Jia (Eds.): PCIC 2024, CCIS 2200, pp. 65–73, 2025.
https://doi.org/10.1007/978-981-97-7812-6_6

However, the practical implementation of causal discovery methods faces significant challenges, especially in terms of evaluation. While randomized controlled trials (RCTs) [14] are the gold standard, they are often difficult to conduct in real-world scenarios. Currently, the evaluation of these methods primarily relies on simulated data and real-world datasets, but the former may not fully capture the complexities of real-world situations. Therefore, there is an urgent need to develop more comprehensive and effective evaluation metrics to accurately assess the performance of causal discovery methods.

In this paper, we proposed a method for evaluating causal discovery in the absence of a ground truth causal graph. Inspired by the deep learning approach of dividing data into training and test sets, we leveraged the fact that the underlying causal graph from the same data source remains unchanged. Therefore, the underlying causal graphs of the training and test sets are consistent. We first used the training set to learn the causal graph. Then, we evaluated the learned causal graph on the test set by performing a Markov blanket test to assess its correctness. Next, we extracted all the edges from the learned causal graph and employed four different methods to determine the causal direction of these edges. Finally, we integrated the results of these methods through weighted voting to obtain the final accuracy of the causal graph evaluation.

2   Preliminaries

2.1   Dataset Splitting Strategy

In deep learning, the proper division of datasets is crucial for evaluating model performance. The training set is used for model training, while the test set is used to verify the model's generalization ability, prevent overfitting, and guide parameter tuning. Our proposed evaluation method focuses on utilizing the invariance of causal structures between training and test sets. By splitting the dataset, causal discovery is performed during the training phase to construct a causal graph, and this causal graph is then used in the test phase to evaluate its accuracy and applicability with the test set data. This method not only validates the model's effectiveness but also provides a solid foundation for subsequent causal inference.

2.2   Markov Blanket

Describes the smallest set of variables in a directed graph model that renders a node conditionally independent of other nodes in the network, given this set [10]. Specifically, the Markov blanket of a node includes its parent nodes, child nodes, and other parent nodes of its child nodes (spouse nodes). In Bayesian networks, given a node's Markov blanket, the node is conditionally independent of all other nodes in the network. Therefore, the Markov blanket has the following property:

Proposition: *With this definition, if X_i is conditionally independent of all other nodes $\forall X_j \in X \setminus (MB(X_i) \cup X_i)$ in the network, given its Markov blanket*

$MB(X_i)$, *then we have:*

$$X_i \perp\!\!\!\perp X_j \mid MB(X_i) \tag{1}$$

In this paper, we propose an evaluation method that first utilizes the causal graph learned from the training set to obtain each node's Markov blanket. Then, based on Eq. (1), we perform the Markov Blanket Test (MBT) on the test set data. For each causal discovery method, we report comprehensive test results: if all nodes pass the MBT, the result is "Satisfy"; otherwise, if any node fails, the result is "Violate".

2.3   Causal Pair Learning Methods

Causal pair learning methods encompass various models designed to identify causal relationships between two variables. These methods leverage information such as functional relationships, noise characteristics, and distribution structures between variables to infer causal directions through different assumptions and algorithms. For bivariate causal identification, it is typically modeled as an additive noise causal model where a cause variable C and an effect variable E are related through a function f, with R representing residuals:

$$E = f(C) + R \tag{2}$$

Causal direction is usually identified based on the functional relationship between the cause and effect variables or asymmetric residuals. The main methods include:

(1) Linear Non-Gaussian Acyclic Model (LiNGAM) [13]: Assumes that causal relationships among system variables are composed of linear functions and a non-Gaussian noise term. LiNGAM determines causal direction by identifying the independence of noise terms and causal variables, using Independent Component Analysis (ICA) to solve for the causal weight matrix.

(2) Additive Noise Model (ANM) [5]: Models the effect variable as an additive function of the cause variable and a noise variable. ANM does not specify the type of additive function or the distribution of the residuals. Typically, the noise is assumed to be a nonlinear function with a Gaussian distribution. For more than two variables, each additive function and residual noise distribution can vary, making the Linear Non-Gaussian Acyclic Model a special case or part of the ANM model.

(3) Information Geometric Causal Inference (IGCI) [3]: Uses the distribution structure of variables and mapping functions to identify causal direction. IGCI assumes that the distribution of the cause variable and the function are independent natural mechanisms. Thus, in the correct causal direction, the variable distribution and the mapping function do not contain information about each other. Causal direction is identified based on Kullback-Leibler (KL) divergence, with smaller KL divergence indicating the causal direction.

(4) Regression Error Causal Inference (RECI) [1]: Infers causal relationships between paired variables by comparing the fitting errors of causal and non-causal models. RECI assumes that functions are invertible and differentiable, variable distributions are within a finite range, and noise is unbiased with unit variance. By calculating the mean squared error in both directions, the direction with the smaller error is chosen as the causal direction.

These methods share a commonality in that they all use different forms of independence assumptions and error comparisons, leveraging functional relationships and noise characteristics among variables to identify causal directions, thus providing a set of effective tools for causal inference in complex systems.

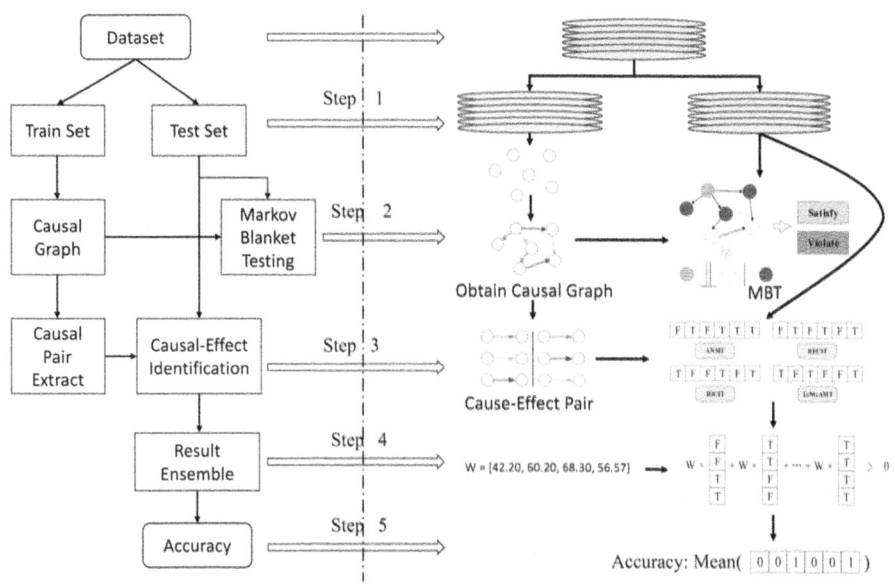

Fig. 1. Our proposed evaluation methods.

3   Methods

As shown in Fig. 1, our proposed evaluation method consists of five steps, detailed as follows:

Step 1: First, we divide the dataset into training and test sets. Unless otherwise specified, we use a split ratio of 7:3.

Step 2: Next, we use the features of the training set as variables and employ causal discovery methods to learn the causal graph. After obtaining the causal graph, we perform a Markov blanket test using the test set. The significance of the Markov blanket test lies in determining the causal relationship between nodes by checking if the Markov blanket of a specific node satisfies certain conditions. If a node's Markov blanket includes all nodes that directly influence it, and the states of these nodes are known, the node can be considered conditionally independent of other nodes. This conditional independence can be interpreted as a causal relationship, indicating causal influence between nodes. Thus, satisfying the Markov blanket test is the fundamental requirement for the causal graph obtained by the causal discovery algorithm. This can be expressed by the following formula:

$$v \perp\!\!\!\perp u | MB(v), \forall v \in V, u \neq v, \forall u \in V \setminus MB(v) \tag{3}$$

where $v \in V$ is the set of nodes and E is the set of edges, we first find its Markov blanket MB. For each node, we perform Markov blanket testing, with the confidence level set at 0.05. For a causal graph G, if all nodes satisfy the Markov blanket test (MBT), we output "Satisfy"; conversely, if any node fails to satisfy MBT, we return "Violate".

Step 3: We extract causal pairs (edges) from the learned causal graph and then use the four causal direction determination methods described in Chap. 3 to judge the direction of each causal pair. Each determination method's result is labeled as True or False, indicating consistency with the causal graph (True for consistent, False for inconsistent). These methods are referred to as LiNGAM Testing, ANM Testing, RECI Testing, and IGCI Testing. For the ANM algorithm, we set its regression algorithm to Gaussian Process Regression (GPR) [19], and for RECI, we set it to linear regression.

Step 4: After obtaining the causal direction judgments for each edge, we form a $\mathcal{R}^{4x|V|}$ judgment matrix. We then perform weighted voting integration on the results. Specifically, we use the accuracy rates of the four causal direction methods on the CauseEffectPair dataset from the literature [17] as the weight vector. We apply these weights to the judgments of each edge, and the final determination result is decided by majority voting through a threshold θ. Unless otherwise specified, the threshold θ is set to 0.6.

Step 5: To obtain the final accuracy of the judgments, we average the judgment vector to get the overall evaluation of the causal graph.

Output: The evaluation output for the causal graph learned by the causal discovery method includes the Markov blanket test result (Satisfy or Violate) and the accuracy of the causal graph learning (Accuracy).

4 Experiments

In the experiments, we use four causal mechanisms from the literature [6] to simulate synthetic causal discovery datasets and evaluate eight different types of causal discovery methods on both synthetic and real datasets (the Sachs dataset).

4.1 Datasets and Baseline Methods

Synthetic Datasets: For the synthetic datasets, we generate samples with a size of 5000 and set the training and test set ratio to 7:3. The four causal mechanisms are linear synthesis, polynomial synthesis, sigmoid mix synthesis, and Gaussian process mix synthesis.

Real Sachs Dataset: The Sachs dataset [12] measures the expression levels of phospholipids and proteins in human cells. It simultaneously measures 11 phospholipids and phosphorylated proteins in thousands of immune system cells. The Sachs dataset is continuous, representing the concentration of the molecules studied. Standard methods in the literature assume that concentrations follow a Gaussian distribution and use Gaussian Bayesian Networks to construct the protein signaling network. The dataset consists of 11 nodes, 17 edges, and 7466 instances, with an average Markov blanket size of 3.09.

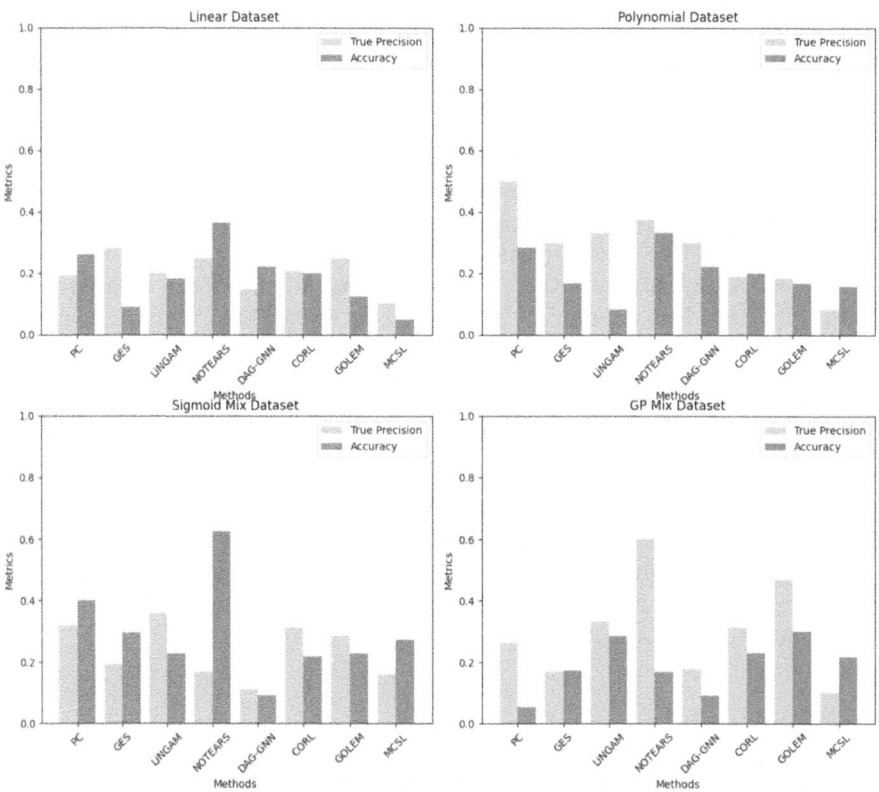

Fig. 2. Experiment on Synthetic datasets.

Baseline Methods: We use the following causal discovery methods: PC [15], GES [2], LiNGAM [13], NOTEARS [21], DAG-GNN [20], GOLEM [8], MCSL [9], and CORL [18].

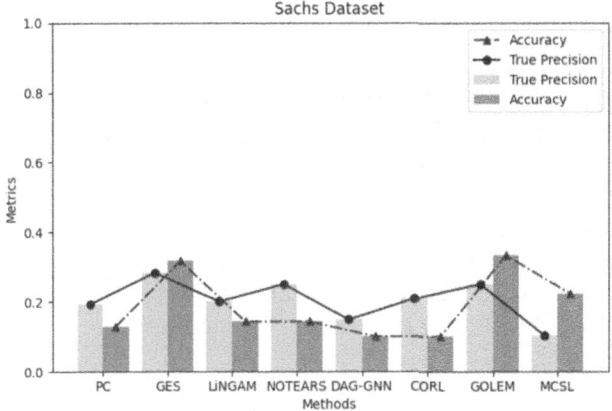

Fig. 3. Experiment on Sachs datasets.

4.2   Result Analysis

To validate the proposed evaluation method, we conduct analyses from two perspectives. The first perspective compares the numerical values of the estimated accuracy with the true accuracy, with closer values indicating better performance. The second perspective examines the trend similarity between the proposed method and the true accuracy across multiple causal discovery methods. It is important to note that all causal graphs learned by the causal discovery methods in the experiments satisfy the Markov blanket tests.

Synthetic Datasets: Figure 2 presents our experimental results on four synthetic datasets. From Fig. 2, we can observe that the accuracy obtained by the proposed method shows minimal numerical difference from the true accuracy in the linear synthesis mechanism, with an average difference within 0.1. In the Gaussian process mechanism and the sigmoid mechanism, the NOTEARS method exhibits significant fluctuations. However, methods of the same type show smaller numerical differences in the sigmoid mechanism and maintain the same trend as the true accuracy in the Gaussian process mechanism. Additionally, methods based on continuous optimization demonstrate conclusions consistent with both evaluation perspectives across all four synthetic datasets.

Real Sachs Dataset: Figure 3 displays the evaluation results of our method on the real dataset. The line chart reveals that the trend of the proposed method's accuracy curve closely follows that of the true accuracy curve, showing similar upward trends and very close numerical values. Therefore, the proposed evaluation method can reflect the performance of causal discovery methods on real data to a certain extent.

5   Conclusion

A method for evaluating causal discovery in the absence of a ground truth causal graph is proposed in this paper. Real-world data is assessed by examining the consistency of the underlying causal graphs in training and test datasets, using an evaluation approach similar to that in deep learning. Markov blanket tests are conducted and the causal direction of each edge for the causal graphs learned from the training set is determined. The experimental results demonstrate that the true accuracy can be reflected to some extent by the proposed method. The focus of future work will be on improving the strategies for determining accuracy, as well as the global and local assessments of causal graphs.

Acknowledgment. This work supported by the National Major Science and Technology Projects of China [grant number 2022YFB3303302], the State Key Laboratory of Complex Electromagnetic Environment Effects on Electronics and Information System [grant number CEMEE2023G0202] and the National Natural Science Foundation of China [grant numbers 62207007].

References

1. Blöbaum, P., Janzing, D., Washio, T., Shimizu, S., Schölkopf, B.: Analysis of cause-effect inference by comparing regression errors. PeerJ. Comput. Sci. **5**, e169 (2019)
2. Chickering, D.M.: Optimal structure identification with greedy search. J. Mach. Learn. Res. **3**, 507–554 (2002)
3. Daniušis, P., et al.: Inferring deterministic causal relations. In: Proceedings of the Twenty-Sixth Conference on Uncertainty in Artificial Intelligence, pp. 143–150 (2010)
4. Deng, L., Liu, Y.: Deep Learning in Natural Language Processing. Springer, Singapore (2018). https://doi.org/10.1007/978-981-10-5209-5
5. Hoyer, P., Janzing, D., Mooij, J.M., Peters, J., Schölkopf, B.: Nonlinear causal discovery with additive noise models. Adv. Neural Inf. Process. Syst. **21** (2008)
6. Kalainathan, D., Goudet, O., Guyon, I., Lopez-Paz, D., Sebag, M.: Structural agnostic modeling: adversarial learning of causal graphs. J. Mach. Learn. Res. **23**(219), 1–62 (2022)
7. Miotto, R., Wang, F., Wang, S., Jiang, X., Dudley, J.T.: Deep learning for healthcare: review, opportunities and challenges. Brief. Bioinform. **19**(6), 1236–1246 (2018)
8. Ng, I., Ghassami, A., Zhang, K.: On the role of sparsity and DAG constraints for learning linear DAGs. Adv. Neural. Inf. Process. Syst. **33**, 17943–17954 (2020)
9. Ng, I., Zhu, S., Fang, Z., Li, H., Chen, Z., Wang, J.: Masked gradient-based causal structure learning. In: Proceedings of the 2022 SIAM International Conference on Data Mining (SDM), pp. 424–432. SIAM (2022)
10. Pearl, J.: Probabilistic Reasoning in Intelligent Systems: Networks of Plausible Inference. Morgan Kaufmann, San Mateo (1988)
11. Pearl, J., Mackenzie, D.: The Book of Why: The New Science of Cause and Effect. Basic Books, New York (2018)
12. Sachs, K., Perez, O., Pe'er, D., Lauffenburger, D.A., Nolan, G.P.: Causal protein-signaling networks derived from multiparameter single-cell data. Science **308**(5721), 523–529 (2005)

13. Shimizu, S., Hoyer, P.O., Hyvärinen, A., Kerminen, A., Jordan, M.: A linear non-gaussian acyclic model for causal discovery. J. Mach. Learn. Res. **7**(10), 2003–2030 (2006)
14. Sibbald, B., Roland, M.: Understanding controlled trials Why are randomised controlled trials important? BMJ Br. Med. J. **316**(7126), 201 (1998)
15. Spirtes, P., Glymour, C., Scheines, R.: Causation, Prediction, and Search. MIT Press, Cambridge (2001)
16. Voulodimos, A., Doulamis, N., Doulamis, A., Protopapadakis, E.: Deep learning for computer vision: a brief review. Comput. Intell. Neurosci. **2018** (2018)
17. Wang, L., Huang, S., Wang, S., Liao, J., Li, T., Liu, L.: A survey of causal discovery based on functional causal model. Eng. Appl. Artif. Intell. **133**, 108258 (2024)
18. Wang, X., et al.: Ordering-based causal discovery with reinforcement learning. arXiv preprint arXiv:2105.06631 (2021)
19. Williams, C.K., Rasmussen, C.E.: Gaussian Processes for Machine Learning, vol. 2. MIT Press, Cambridge (2006)
20. Yu, Y., Chen, J., Gao, T., Yu, M.: DAG-GNN: DAG structure learning with graph neural networks. In: International Conference on Machine Learning, pp. 7154–7163. PMLR (2019)
21. Zheng, X., Aragam, B., Ravikumar, P.K., Xing, E.P.: DAGs with no TEARS: continuous optimization for structure learning. Adv. Neural. Inf. Process. Syst. **31**, 9492–9503 (2018)

Optimizing Experimental Design for Causal Effect Estimation with Partial Measurements

Leopold Mareis[1,2]([envelope]) [ORCID]

[1] Fraunhofer Institute for Cognitive Systems IKS, Hansastraße 32, Munich, Germany
[2] TUM School of Computation, Information and Technology,
Technical University of Munich, Munich, Germany
`leopold.mareis@tum.de`

Abstract. Instrumental variable regression quantifies causal effects between a possibly confounded treatment variable X_2 and a response variable X_3 by leveraging an instrument X_1. Our work considers the setting where some prior information of the joint distribution of X_{123} is given, potentially through an initial dataset. However, further samples must be gathered to improve the accuracy of the estimation. We show that under specific parameter configurations in a Gaussian graphical model, taking partial samples from, e.g., X_{12} can reduce the asymptotic variance of a consistent estimator. This idea is developed by adding a budget constraint over the cost per (partial) sample. The optimization problem is analytically solvable over the real numbers and gives the optimal number of requested partial and complete samples. We provide significance level, power, and sample-size calculations for detecting a non-zero causal effect under optimal budget allocation. Our method can considerably reduce the necessary budget and the number of complete samples. Finally, we showcase the advantages and applicability of adaptive causal effect estimation for automotive analytics and pharmaceutical research.

Keywords: Causal Effect Estimation · Experimental Design · Graphical Model

1 Introduction

The problem of forming decisions under uncertainty is central to the success of humans and entities. If done correctly, the subsequent action of a decision should have a certain effect on the outcome of interest. A common statistical tool to represent and specify the relationships between random variables is a directed acyclic graph (DAG). It encodes whether direct effects between variables exist and provides a powerful machinery when combined with density estimates. Estimating causal effects between two variables is essential for understanding the random system, and these finite-sample estimates are uncertain. In this paper, we study the problem of gathering additional data to increase the accuracy of causal

© The Author(s), under exclusive license to Springer Nature Singapore Pte Ltd. 2025
X.-H. Zhou and J. Jia (Eds.): PCIC 2024, CCIS 2200, pp. 74–85, 2025.
https://doi.org/10.1007/978-981-97-7812-6_7

effect estimates. Inspired by applications, we allow for requesting non-complete, and thus cheaper, observations. Depending on the domain, cheaper can also mean faster or more ethical. Our approach leverages asymptotic variances in a Gaussian Graphical Model [16] to form decisions on subsequent data requests. Although this approach can be applied to arbitrary DAGs, we focus on the well-studied instrumental variable (IV) approach [1,10,14]. If the requirements are satisfied, this setup allows for causal effects estimation without the necessity of randomization. To make informed decisions, our theory provides information on the interplay between available budget, p-values, and power computations. This is especially important in safety-critical domains where data sampling is restricted and needs to be approved and assessed. Most approaches for adaptive data sampling focus on the non-predetermined number of samples and stop the data gathering procedure when pre-specified thresholds are met [11,12]. In non-full-information estimation, active learning methods want to uncover the label of samples leading to a high information gain [2,13]. And finally, [6] selected statically covariates in valid adjustment sets based on asymptotic variances.

The paper is structured as follows. Section 3 studies the instrumental variable models where more partial instrument-treatment data is helpful. Section 4 provides concrete optimized plans for requesting data under a limited budget. This theory is extended in Sect. 5 to detect positive effects via hypothesis testing before two real applications follow in Sect. 6.

2   Setup

Model 1. *We consider the causal model given by the structural equation model*

$$\begin{pmatrix} X_1 \\ X_2 \\ X_3 \end{pmatrix} = \begin{pmatrix} 0 & 0 & 0 \\ \lambda_{12} & 0 & 0 \\ 0 & \lambda_{23} & 0 \end{pmatrix} \begin{pmatrix} X_1 \\ X_2 \\ X_3 \end{pmatrix} + \varepsilon, \ \varepsilon \sim N \left(\begin{pmatrix} 0 \\ 0 \\ 0 \end{pmatrix}, \begin{pmatrix} \omega_{11} & 0 & 0 \\ 0 & \omega_{22} & \omega_{23} \\ 0 & \omega_{23} & \omega_{33} \end{pmatrix} \right),$$

with $\lambda_{12} \neq 0, \lambda_{23}, \omega_{23} \in \mathbb{R}$ and $\omega_{11}, \omega_{22}, \omega_{33} > 0$ satisfying $\omega_{22}\omega_{33} > \omega_{23}^2$.

The variable X_1 represents the instrument, the variable X_2 represents the treatment, and the variable X_3 represents the outcome. Instead of explicitly modeling a confounding variable, we represent the hidden relationship between X_2 and X_3 by ω_{23}. This is equivalent under the joint linearity assumption, and if $\omega_{23} \neq 0$, confounding is present. Model 1 can also be written in matrix notation as $X = \Lambda^{\top} X + \varepsilon$ where $\varepsilon \sim N(0, \Omega)$. As X follows a multivariate normal distribution, its variance is given by

$$\text{Var}(X) = \begin{pmatrix} \omega_{11} & \omega_{11}\lambda_{12} & \omega_{11}\lambda_{12}\lambda_{23} \\ \cdot & \omega_{22} + \omega_{11}\lambda_{12}^2 & \omega_{23} + \lambda_{23}\sigma_{22} \\ \cdot & \cdot & \omega_{33} + 2\omega_{23}\lambda_{23} + \lambda_{23}^2\sigma_{22} \end{pmatrix} =: \Sigma. \tag{1}$$

Under this setup, the parameter of interest giving the causal effect of X_2 on X_3 is λ_{23}. If there exists a connection between the instrument and the treatment, i.e. $\lambda_{12} \neq 0$, then $\hat{\lambda}_{23} = \frac{\hat{\sigma}_{13}}{\hat{\sigma}_{12}}$ is an unbiased estimator [3]. This estimator has two properties. First, it is solely based on entries of the estimated covariance matrix $\hat{\Sigma}$, and the numerator and denominator can be estimated on distinct or shared datasets. Secondly, it is unstable when the denominator is close to 0, so if $|\omega_{11}\lambda_{12}|$ is small – in this case we say X_1 is a *weak instrument*. Under the weak instrument scenario, our strategy is to measure more data from X_{12} to reduce the variance of $\hat{\sigma}_{12}$ and be more confident in its estimate, motivating the next section.

3   More Instrument Data

We assume to have $n_1 \in \mathbb{N}$ initial datapoints of X_{123} and additionally n_2 datapoints of X_{12} which were measured to further investigate the relationship between the instrument and the treatment variable. With the available information, the estimand $\hat{\lambda}_{12}$ can be computed based on n_1 samples for $\hat{\sigma}_{13}$ and $n_1 + n_2$ samples for $\hat{\sigma}_{12}$. The following theorem derives its asymptotic variance in terms of n_1 with Cramér's Theorem, and the full proof can be found in Appendix 8.

Theorem 1. *Let X follow Model 1. Let $\hat{m}_{jk;1} = \frac{1}{n_1} \sum_{i=1}^{n_1} X_j^{(i)} X_k^{(i)}$ and $\hat{m}_{jk;2} = \frac{1}{n_2} \sum_{i=n_1+1}^{n_1+n_2} X_j^{(i)} X_k^{(i)}$ be the sample moments of $X_j X_k$ on the two datasets. Furthermore, let the asymptotic relation n_2/n_1 be γ. The sample size adjusted estimator for λ_{23} is*

$$\hat{\lambda}_{23} = \frac{\hat{m}_{13;1}}{\frac{1}{1+\gamma}\hat{m}_{12;1} + \frac{\gamma}{1+\gamma}\hat{m}_{12;2}}$$

and it attains the asymptotic distribution $\sqrt{n_1}(\hat{\lambda}_{23} - \lambda_{23}) \xrightarrow{d} N(0, v(\gamma))$ with the variance

$$v(\gamma) = \frac{1}{\sigma_{12}^2}\left(\frac{\sigma_{13}^2\sigma_{11}\sigma_{22}}{(1+\gamma)\sigma_{12}^2} - 2\frac{\sigma_{13}\sigma_{11}\sigma_{23}}{(1+\gamma)\sigma_{12}} + \sigma_{11}\sigma_{33} + 2\sigma_{13}^2 \right) \tag{2}$$

$$= \frac{1}{\omega_{11}\lambda_{12}^2}\left(\frac{\gamma}{(1+\gamma)}\lambda_{23}^2\omega_{22} + \frac{(2+3\gamma)}{(1+\gamma)}\lambda_{23}^2\omega_{11}\lambda_{12}^2 + \frac{2\gamma}{(1+\gamma)}\lambda_{23}\omega_{23} + \omega_{33} \right).$$

The variance $v(\gamma)$ in original parameterization suggests that increasing ω_{22} or ω_{33} increases the uncertainty of $\hat{\lambda}_{23}$ in all configurations. For the remaining original parameters, the variance can decrease or increase, mainly depending on the sign and magnitude of $\omega_{23}\lambda_{23}$. Note that $\omega_{22}\omega_{33} > \omega_{23}^2$ prevents negative $v(\gamma)$ if $\omega_{23}\lambda_{23}$ approaches $-\infty$. In the special case where only the primary dataset is available, so $\gamma = n_2 = 0$, Theorem 1 reduces as follows:

Corollary 1. *Under the assumptions of Theorem 1 with $\gamma = 0$, the estimator $\hat{\lambda}_{23}$ attains the following asymptotic distribution:*

$$\sqrt{n}(\hat{\lambda}_{23} - \lambda_{23}) \xrightarrow{d} N(0, 2\lambda_{23}^2 + \frac{\omega_{33}}{\omega_{11}\lambda_{12}^2})$$

In the scenario with one available dataset on X_{123}, increasing λ_{23}^2 or ω_{33} or decreasing ω_{11} or λ_{12}^2, increases the variance. Intuitively, this means that the weaker the instrument is, the worse the estimand $\hat{\lambda}_{23}$ behaves. In the latter case, it can, therefore, be sensible to gather more data on X_{12}:

Corollary 2. *Let X follow Model 1. The asymptotic variance function $v(\gamma)$ is decreasing for $\gamma > 0$ if and only if*

$$\lambda_{12}^2 < -\frac{\omega_{22} + 2\omega_{23}/\lambda_{23}}{\omega_{11}}. \tag{3}$$

In particular, this can only be the case if $\omega_{23}\lambda_{23} < 0$.

This statement must be understood in the context of the chosen asymptotics, where only the number of X_{123} samples is in focus. It treats n_1 samples of X_{123} equally as n_1 samples of X_{123} with additional n_2 samples of X_{12}. Consequently, measuring only X_{123} might not be the variance-reducing strategy as further X_{12} samples can be requested without penalization. Particularly, in cases where the sign of ω_{23}/λ_{23} is negative, so in cases with counteracting effect paths between X_2 and X_3, Corollary 2 implies that more X_{12} data is beneficial. Additionally, Eq. 3 implies that small $\lambda_{12}^2\omega_{11}$, so a weak instrument scenario, is necessary for additional partial samples. As an extension and to leverage the reduced cost of measuring only X_{12}, we now add a budget to formulate the problem of informed future data acquisition.

4   Gathering Further Samples Under a Fixed Budget

Suppose that under Model 1, n samples of X_{123} are collected. Further, let there be an additional budget of $b \in \mathbb{R}$ from which samples of X_{123} and X_{12} can be requested for a cost of c_1, respectively c_2. These costs ought to satisfy the relation $c_2 < c_1$. The goal is to spend the available budget in an asymptotically optimal way to reduce the variance of the estimator $\hat{\lambda}_{12}$. Let m_1 denote the number of additional X_{123} samples and m_2 the number of additional X_{12} samples. With the variance function v from Eq. (2), we obtain the optimization problem

$$\min_{(m_1,m_2)\in\mathbb{N}^2} \frac{v(\frac{m_2}{n+m_1})}{n + m_1} \quad , \text{ s.t. } \quad c_1 m_1 + c_2 m_2 \leq b. \tag{4}$$

where the objective is derived from the asymptotic variance of $\hat{\lambda}_{23} - \lambda_{23}$, scaled by the number of complete cases samples X_{123}. We denote a solution by (m_1^*, m_2^*).

Theorem 2. *Let X follow Model 1, and let $b, c_1, c_2 > 0$ as well as $n \in \mathbb{N}$ set the Optimization Problem (4). The solution to the related continuous optimization problem is given by $(0, b/c_2)$ if*

$$\chi_1 := (c_1 - c_2)(\sigma_{11}\sigma_{13}^2\sigma_{22} - 2\,\sigma_{11}\sigma_{12}\sigma_{13}\sigma_{23}) \leq 2\,\sigma_{12}^2\sigma_{13}^2 c_2 + \sigma_{11}\sigma_{12}^2 c_2\sigma_{33} =: \chi_2.$$

Otherwise, a positive amount of X_{12} samples are requested and the solution reads

$$\Big(\frac{b - \xi c_2 n}{c_1 + \xi c_2}, \frac{\xi(b + c_1 n)}{c_1 + \xi c_2}\Big), \quad \text{where } \xi = \min\Big(-1 + \frac{\sqrt{\chi_1\chi_2}}{\chi_2}, \frac{b}{c_2 n}\Big).$$

The condition $\chi_1 \leq \chi_2$ expressed in original parameterization is

$$\frac{c_1}{c_2}(-2\lambda_{23}\omega_{23} - \lambda_{23}^2\omega_{22} - \lambda_{23}^2\omega_{11}\lambda_{12}^2) \leq 2\,\omega_{11}\lambda_{12}^2\lambda_{23}^2 + \omega_{33}$$

and implies three conceptual findings: Firstly, as expected by Corollary 1, additional X_{12} samples are only beneficial if the causal effect λ_{23} and the confounding strength ω_{23} have opposite signs. Secondly, only the model parameters and the ratio c_1/c_2 determine whether additional partial samples are requested. And thirdly, weakening the instrument λ_{12}^2 always weakens the condition $\chi_1 < \chi_2$ condition in favour of more potential X_{12} samples.

Example 1. Let X follow Model 1 with $\omega_{11} = 0.5$, $\omega_{22} = 1$, $\omega_{33} = 2$, $\lambda_{12} = 0.2$, $\lambda_{23} = 1$ and $\omega_{23} = -1$. In this case, the covariance between the treatment and outcome is 0.1, underestimating the actual causal effect of λ_{23} by a factor of 10. Corollary 2 indicates a benefit in an increased number of X_{12} samples. Let us further assume that $n = 500$ samples were gathered from X_{123} and that there is an additional budget of $b = 250$ from which samples of X_{123} can be requested for $c_1 = 3$ or from X_{12} for $c_2 = 1$. The floored optimal data request is $(m_1, m_2) \approx (20, 187)$ giving an approximate variance of $\approx 17.12\%$ of the estimator $\hat{\lambda}_{23}$. In contrast, requesting only samples of X_{123} gives $\approx 17.13\%$ variance. It turns out that 14.8% more budget would be necessary to match the partial measurements estimator's variance.

5   Sample Size Calculation, Significance Level and Power

When gathering valuable, sensitive, or hazardous data, sample size calculations must often be provided in experimental planning. This ensures enough data is collected for scientific conclusions while minimizing the risk to test subjects and resources. For this, we consider hypothesis tests of the form $H_0 : \lambda_{23} \leq 0$ against $H_A : \lambda_{23} > 0$ to detect wlog a positive effect. The test family has a rejection criterion of the form $\hat{\lambda}_{23} > t$. To represent the uncertainty in the parameter λ, let $v(\gamma; \mu)$ denote the asymptotic variance function obtained by solving the Optimization Problem (4) where the underlying Model 1 satisfies $\lambda_{23} = \mu \in \mathbb{R}$.

Lemma 1 (Significance Level). *Let X follow Model 1. The hypothesis test $H_0 : \lambda_{23} \leq 0$ against $H_A : \lambda_{23} > 0$ with the rejection criterion*

$$\hat{\lambda}_{23} > \sup_{\mu \leq 0} F^{-1}_{N(\mu, v(\gamma^*; \mu)/(n+m_1^*))}(1 - \alpha) =: \tau \qquad (5)$$

then attains the asymptotic significance level $\alpha \in (0, 1)$.

To enforce $\lambda_{23} = \mu$ in practical calculations, use the estimate $\hat{\Sigma}$ to identify the underlying parameters[1] of Λ and Ω, and rebuild the model with $\lambda_{23} = \mu$.

---

[1] $\omega_{11} = \sigma_{11}$, $\lambda_{12} = \sigma_{12}/\sigma 11$, $\omega_{22} = -\sigma_{22} - \omega_{11}\lambda_{12}^2$, $\omega_{23} = \sigma_{23} - \sigma_{22}\sigma_{13}/\sigma_{12}$, $\omega_{33} = \sigma_{33} - 2\omega_{23}\sigma_{13}/\sigma_{12} - \sigma_{13}^2/\sigma_{12}^2 * \sigma_{22}$.

Corollary 3 (Power). *Let X follow Model 1. Let a hypothesis test with threshold τ and significance level $\alpha \in (0,1)$ be given according to Lemma 1. The asymptotic power of that test at effect level $\mu > 0$ is*

$$1 - F_{N(\mu, v(\gamma^*; \mu)/(n+m_1^*))}(\tau). \tag{6}$$

We recommend analyzing the power of effect levels μ around the current estimate $\hat{\lambda}_{23}$ to cover pessimistic and optimistic scenarios.

Corollary 4 (p-value). *Let X follow Model 1. Let $\hat{\lambda}_{23}$ be an estimate of λ_{23} on $n + m_1$ samples of X_{123} and m_2 samples of X_{12}. The p-value of the hypothesis test family from Lemma 1 is*

$$\max_{\mu \leq 0} 1 - F_{N(\mu, v(m_2/(n+m_1); \mu)/(n+m_1))}(\hat{\lambda}_{23}). \tag{7}$$

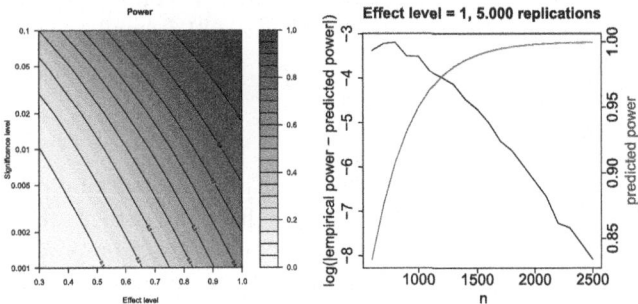

Fig. 1. Example 2: Asymptotic power computations and simulations based on Corollary 3. Note that both y-axes are logarithmic.

Example 2. Figure 1 shows power computations under the setup of Example 1. The left panel visualizes the power contour lines for combinations of the significance level (y-axis) and the reference effect level of the alternative (x-axis). At the reference effect $\lambda_{23} > 0.75$, and significance level $\alpha = 0.05$, the hypothesis test's power already lies above 0.8. In the right panel, the underlying hypothesis tests' at several sample sizes n is given in blue according to Corollary 3. The black line shows the log distance to the empirical power based on 5000 replications. Under the normality assumptions, the asymptotic approximation of the estimator λ_{23} words well.

6   Real Datasets

Observed Confounder. An extended instrumental variable model considers an exogenous additional variable X_4 influencing X_{123}. In this scenario, the conditional covariance matrix $Cov(X_{123}|X_4)$ is equal to the unconditional one from Eq. (1). If additional exogenous information is available, using the conditional covariance matrix through a preprocessing step allows the transfer of previously introduced theory.

6.1 Application Aggressive Driving

Aggressive driving is a risk for drivers and positively affects the number of accidents. In this dataset, 10932 drivers were monitored while driving. Environmental variables such as lighting, temperature, humidity, and car-specific variables were recorded. The causal DAG obtained by the PC-Algorithm [9] relates the variables and is displayed in Fig. 2 in Appendix 9. This DAG suggests an instrumental variable setup with $(X_1, X_2, X_3) =$ (lighting, driving style, speed). Therefore, the question of interest is whether the driving style influences the car's average speed. Here, the direct effect is -5.6 km/h average speed per unit increase of aggressive driving style. Hence, actions to relax the driving style should be taken to reduce the average speed, preventing unnecessary accidents [15]. When applying our Adaptive Sampling Method (4), the competing estimates have the following variances: Optimized variance: 1.012. Baseline variance: 1.274. Theoretical optimized variance: 7.329. Theoretical baseline variance: 7.526. Our proposed optimized method achieves a 21% smaller variance over the baseline procedure.

6.2 ICU Data

In the intensive care unit, the water level of patients must be controlled, and Lasix (here Furosemide) is used to drain urine from the patients. Reasons can be malfunctioning kidneys or excess water due to infusions. A preprocessed MIMIC-dataset [8] on 2000 patients represents an instrumental variable scenario by construction. A patient-wide, caregiver-specific average dosage is used as the instrument; the actual given amount acts as X_2 and the drained water in ml/min as X_3. When applying our Adaptive Sampling Method (4), the competing estimates have the following variances: Optimized variance: $8.879e - 09$. Baseline variance: $2.809e - 08$. Theoretical optimized variance: $1.098e - 07$. Theoretical baseline variance: $1.1038e - 07$. Our proposed optimized method achieves 68% smaller variance over the baseline procedure.

7 Conclusion

We have proposed a framework to plan data requests with partial measurements for a given cost function. It is a conceptually different approach to prior adaptive experimental design methods; however, it keeps the same intention- to reduce or shift the number of necessary total samples. Theorem 2 gives a criterion when partial measurements are helpful and explicitly states the optimal future data request. When there is no need for partial data, it can indicate aligned confounding and causal effect signs in the underlying model. The central idea of Optimization Problem 4, combining asymptotic variances and budgeting, is not limited to instrumental variable problems and can be derived for more complex causal DAGs. This holds under the premise of identifiable target effects. Even relaxing the Gaussian Graphical Model assumption and comparing the asymptotic normality of maximum likelihood estimators on partial datasets is feasible.

Acknowledgments. This work has been funded by the Federal Ministry of Education and Research (BMBF) as part of AutoDevSafeOps (01IS22087Q) ant the Bavarian Ministry for Economic Affairs, Regional Development and Energy as part of a project to support the thematic development of the Fraunhofer Institute for Cognitive Systems.

Disclosure of Interests. The authors have no competing interests to declare that are relevant to the content of this article.

8   Appendix A: Proofs and Derivations

In this appendix, we collect the proofs from the main document.

Proof of Theorem 1

Proof. By scaling the asymptotic of $\hat{m}_{12;2}$ by $1/\sqrt{\gamma}$, we obtain

$$\sqrt{n_1}\left(\begin{pmatrix}\hat{m}_{12;1}\\\hat{m}_{13;1}\\\hat{m}_{12;2}\end{pmatrix} - \begin{pmatrix}\sigma_{12}\\\sigma_{13}\\\sigma_{12}\end{pmatrix}\right) \xrightarrow{d} N\left(0, \begin{pmatrix}Var(X_1X_2) & Cov(X_1X_2,X_1X_3) & 0\\ \cdot & Var(X_1X_3) & 0 \\ \cdot & \cdot & Var(X_1X_2)/\gamma\end{pmatrix}\right).$$

As X is centered, the entries of the asymptotic covariance matrix have the form $\mathbb{E}[X_i^2 X_j^2] = \sigma_{ii}\sigma_{jj} + 2\sigma_{ij}^2$ and $\mathbb{E}[X_i^2 X_j X_k] = \sigma_{ii}\sigma_{jk} + 2\sigma_{ij}\sigma_{ik}$ (see [7]). With

$$(\hat{m}_{12;1}, \hat{m}_{13;1}, \hat{m}_{12;2}) \mapsto (1+\gamma)\hat{m}_{13;1}/(\hat{m}_{12;1} + \gamma\hat{m}_{12;2}) = \hat{\lambda}_{12},$$

the asymptotic distribution of $\hat{\lambda}_{12}$ must be, by Cramér's Theorem (see, e.g., [4], Thm. 7), Gaussian and its variance can be computed with basic algebra as

$$\left(\frac{-\sigma_{13}}{(1+\gamma)\sigma_{12}^2}, \sigma_{12}^{-1}, \frac{-\gamma\sigma_{13}}{(1+\gamma)\sigma_{12}^2}\right)\begin{pmatrix}Var(X_1X_2) & Cov(X_1X_2,X_1X_3) & 0\\ \cdot & Var(X_1X_3) & 0 \\ \cdot & \cdot & Var(X_1X_2)/\gamma\end{pmatrix}\begin{pmatrix}\frac{-\sigma_{13}}{(1+\gamma)\sigma_{12}^2}\\ \sigma_{12}^{-1} \\ \frac{-\gamma\sigma_{13}}{(1+\gamma)\sigma_{12}^2}\end{pmatrix}$$

$$= \frac{\sigma_{13}^2}{(1+\gamma)^2\sigma_{12}^4}\mathbb{E}[(X_1X_2)^2] - 2\frac{\sigma_{13}}{(1+\gamma)\sigma_{12}^3}Cov(X_1X_2,X_1X_3)$$
$$+ \frac{1}{\sigma_{12}^2}\mathbb{E}[X_1X_3] + \frac{\gamma\sigma_{13}^2}{(1+\gamma)^2\sigma_{12}^4}\mathbb{E}[(X_1X_2)^2]$$

$$= \frac{\sigma_{13}^2}{(1+\gamma)^2\sigma_{12}^4}\left(\sigma_{11}\sigma_{22} + 2\sigma_{12}^2\right) - 2\frac{\sigma_{13}}{(1+\gamma)\sigma_{12}^3}\left(\sigma_{11}\sigma_{23} + \sigma_{12}\sigma_{13}\right)$$
$$+ \frac{1}{\sigma_{12}^2}\left(\sigma_{11}\sigma_{33} + 2\sigma_{13}^2\right) + \frac{\gamma\sigma_{13}^2}{(1+\gamma)^2\sigma_{12}^4}\left(\sigma_{11}\sigma_{22} + 2\sigma_{12}^2\right)$$

$$= \frac{1}{\sigma_{12}^2}\left(\frac{\sigma_{13}^2}{(1+\gamma)\sigma_{12}^2}\left(\sigma_{11}\sigma_{22} + 2\sigma_{12}^2\right) - 2\frac{\sigma_{13}}{(1+\gamma)\sigma_{12}}\left(\sigma_{11}\sigma_{23} + \sigma_{12}\sigma_{13}\right) + \left(\sigma_{11}\sigma_{33} + 2\sigma_{13}^2\right)\right)$$

$$= \frac{1}{\sigma_{12}^2}\left(\frac{\sigma_{13}^2\sigma_{11}\sigma_{22}}{(1+\gamma)\sigma_{12}^2} - 2\frac{\sigma_{13}\sigma_{11}\sigma_{23}}{(1+\gamma)\sigma_{12}} + \sigma_{11}\sigma_{33} + 2\sigma_{13}^2\right).$$

This representation of the variance as a function of Σ can now be expressed in the original parameterization of Λ and Ω. For this, we use Eq. (1) and find with basic algebra

$$\frac{1}{\omega_{11}^2\lambda_{12}^2}\left(\frac{\omega_{11}^2\lambda_{12}^2\lambda_{23}^2\omega_{11}(\omega_{22}+\omega_{11}\lambda_{12}^2)}{(1+\gamma)\omega_{11}^2\lambda_{12}^2}-2\frac{\omega_{11}\lambda_{12}\lambda_{23}\omega_{11}(\omega_{23}+\lambda_{23}(\omega_{22}+\omega_{11}\lambda_{12}^2))}{(1+\gamma)\omega_{11}\lambda_{12}}\right.$$

$$\left.+\omega_{11}(\omega_{33}+2\omega_{23}\lambda_{23}+\lambda_{23}^2(\omega_{22}+\omega_{11}\lambda_{12}^2))+2\omega_{11}^2\lambda_{12}^2\lambda_{23}^2\right)$$

$$=\frac{1}{\omega_{11}\lambda_{12}^2}\left(\frac{\lambda_{23}^2\omega_{22}+\lambda_{23}^2\omega_{11}\lambda_{12}^2}{(1+\gamma)}-2\frac{\lambda_{23}\omega_{23}+\lambda_{23}^2\omega_{22}+\lambda_{23}^2\omega_{11}\lambda_{12}^2)}{(1+\gamma)}\right.$$

$$\left.+\omega_{33}+2\omega_{23}\lambda_{23}+\lambda_{23}^2\omega_{22}+3\omega_{11}\lambda_{12}^2\lambda_{23}^2\right)$$

$$=\frac{1}{\omega_{11}\lambda_{12}^2}\left(\frac{\gamma}{(1+\gamma)}\lambda_{23}^2\omega_{22}+\frac{(2+3\gamma)}{(1+\gamma)}\lambda_{23}^2\omega_{11}\lambda_{12}^2+\frac{2\gamma}{(1+\gamma)}\lambda_{23}\omega_{23}+\omega_{33}\right)$$

Proof of Corollary 2

Proof. Taking the derivative gives

$$v'(\gamma)=\frac{1}{(1+\gamma)^2}\frac{1}{\omega_{11}\lambda_{12}^2}\left(\lambda_{23}^2\omega_{22}+\lambda_{23}^2\omega_{11}\lambda_{12}^2+2\lambda_{23}\omega_{23}\right)<0$$

$$\Longleftrightarrow\quad\omega_{22}+\omega_{11}\lambda_{12}^2+2\omega_{23}/\lambda_{23}<0$$

$$\Longleftrightarrow\quad\lambda_{12}^2<-\frac{\omega_{22}+2\omega_{23}/\lambda_{23}}{\omega_{11}}.$$

Proof of Theorem. 2 We focus on the Optimization Problem (4) with the continuous constraint $(m_1, m_2) \in [0, \infty)^2$, so

$$\min_{(m_1,m_2)\in[0,\infty)^2}\frac{v(\frac{m_2}{n+m_1})}{n+m_1}\quad,\text{ s.t. }\quad c_1m_1+c_2m_2\le b.\tag{8}$$

Proof. For a fixed ratio $m_2/(n + m_1)$, more samples will always decrease the objective as $v(m_2/(n + m_1))$ is constant and the denominator increases. Therefore, only the boundary has to be checked, and we can require $c_1m_1 + c_2m_2 = b$. With the relation $m_2 = (b - c_1m_1)/c_2$, the problem

$$\min_{m\in[0,b/c_1]}\frac{v(\frac{1}{c_2}\left(\frac{b-c_1m}{n+m}\right))}{n+m}=:h(c_1,c_2,b,n,m),\tag{9}$$

can be solved equivalently. The solution to the original problem is then given by $(m^*, (b - c_1m^*)/c_2)$. We can equate the derivative $\partial h(c_1, c_2, b, n, m)/\partial m$ to 0 and find

$$\frac{v'(\frac{1}{c_2}\left(\frac{b-c_1m}{n+m}\right))\frac{1}{c_2}\frac{-c_1(n+m)-(b-c_1m)}{(n+m)^2}(n+m)-v(\frac{1}{c_2}\left(\frac{b-c_1m}{n+m}\right))}{(n+m)^2}\overset{!}{=}0$$

$$\Leftrightarrow v'(\xi)\left(\frac{c_1}{c_2}+\xi\right)+v(\xi)=0\quad,\text{ where }\quad\xi=\frac{1}{c_2}\left(\frac{b-c_1m}{n+m}\right).$$

Let's define $\chi_1 := (c_1 - c_2)(\sigma_{11}\sigma_{13}^2\sigma_{22} - 2\,\sigma_{11}\sigma_{12}\sigma_{13}\sigma_{23})$ and $\chi_2 := 2\,\sigma_{12}^2\sigma_{13}^2c_2 + \sigma_{11}\sigma_{12}^2c_2\sigma_{33}$. Then, the necessary equation simplifies to

$$v'(\xi)\left(\frac{c_1}{c_2} + \xi\right) + v(\xi) = \frac{\chi_2\xi^2}{\sigma_{12}^4c_2(1+\xi)^2} + \frac{2\,\chi_2\xi}{\sigma_{12}^4c_2(1+\xi)^2} + \frac{\chi_2 - \chi_1}{\sigma_{12}^4c_2(1+\xi)^2}$$

with solutions $-1 \pm \sqrt{\chi_1\chi_2}/\chi_2$. So $\xi > 0$, which is necessary for $m^* > 0$, is true if and only if $[(c_1 - c_2)/c_2](\sigma_{13}^2\sigma_{22} - 2\,\sigma_{12}\sigma_{13}\sigma_{23}) > 2\,\sigma_{12}^2\sigma_{13}^2/\sigma_{11} + \sigma_{12}^2\sigma_{33}$, or equivalently $(c_1/c_2)(-2\lambda_{23}\omega_{23} - \lambda_{23}^2\omega_{22} - \lambda_{23}^2\omega_{11}\lambda_{12}^2)) > 2\,\omega_{11}\lambda_{12}^2\lambda_{23}^2 + \omega_{33}$.

Proof of Lemma 1

Proof. Let $\mu \le 0$ be an arbitrary parameter in the null hypothesis for the model parameter λ_{23}. Following the recommendation of the Optimization Problem (4), we have that $\hat{\lambda}_{23}$ is asymptotically $N(\mu, v(\gamma;\mu)/(n+m_1))$ distributed. Thus $\lim_{n\to\infty} P_{\lambda_{23}=\mu}(\hat{\lambda}_{23} > F_{N(\mu,v(\gamma;\mu)/(n+m_1))}^{-1}(1-\alpha)) = \alpha$ and the asymptotic significance level is given by

$$\lim_{n\to\infty} P_{H_0}(\text{reject } H_0) = \lim_{n\to\infty} \sup_{\mu\le 0} P_{\lambda_{23}=\mu}\left(\hat{\lambda}_{23} > \tau\right)$$

$$\le \sup_{\mu\le 0} \lim_{n\to\infty} P_{\lambda_{23}=\mu}\left(\hat{\lambda}_{23} > F_{N(\mu,v(\gamma;\mu)/(n+m_1))}^{-1}(1-\alpha)\right) = \alpha.$$

9   Appendix B: Applications – Further Details

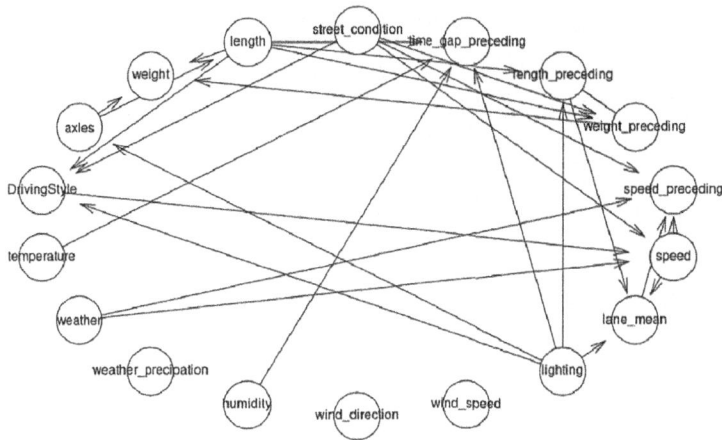

Fig. 2. This DAG on the Aggressive Driving was estimated by the PC-Algorithm.

Aggressive Driving. All measured variables from the publicly available aggressive driving dataset[2] are related in the learned DAG in Fig. 2. The measurement 'driving style' has three levels: *aggressive, normal* and *relaxed* driving. For the adaptive data collection, we assumed 500 initial data samples on X_{123}, an additional budget of 300, and a cost per sample of 1 for X_{123} and 0.3 for X_{12}. The optimized allocation is $m = (211, 293)$. The variances were estimated using subsampling on the dataset. The estimated covariance matrix over the full dataset has the following entries:

$$
\begin{array}{c}
\text{Lighting} \\
\text{Driving Style} \\
\text{Speed}
\end{array}
\begin{pmatrix}
7.436e - 01 & 6.806e - 02 & 2.225e - 01 \\
6.806 \times 10^{-2} & 5.112e - 01 & -8.365e + 00 \\
2.225e - 01 & -8.365e + 00 & 3.531e + 01
\end{pmatrix}
$$

ICU Medication. 20442 patients were treated in the MIMIC-IV ICU database with Lasix, where foley output (urine) was measured during the subsequent 6 h. To eliminate dependencies between samples, we selected every patient's first ICU stay. As the response variable X_3, we selected the *urine/minutes passed* ratio of the last measurement within these 6 h. We selected 2000 patients at random and used their caregiver ID to generate an instrumental caregiver attribute over the remaining 18442 patients. This was the average prescribed amount of the caregiver in the secondary dataset. The estimated covariance matrix over the full dataset has the following entries:

$$
\begin{array}{c}
\text{Caregiver Rate} \\
\text{Medication} \\
\text{Foley Rate}
\end{array}
\begin{pmatrix}
1.345e + 04 & 7.921e + 03 & 0.2424 \\
7.921e + 03 & 1.125e + 05 & 1.038 \\
2.424e - 01 & 1.037 & 0.1082
\end{pmatrix}
$$

For the adaptive data collection, we assumed 200 initial data samples on X_{123}, an additional budget of 100, and a cost per sample of 1 for X_{123} and 1/7 for X_{12}. The optimized allocation is $m = (88, 77)$ with an effect size of $3.064e - 5$. The variances were estimated using subsampling on the remaining dataset.

Assembly Line. In the previous two scenarios, our method advised gathering more partial data. A real-life covariance matrix where this is not the case is

$$
\begin{pmatrix}
7.278e - 10 & -1.451e - 07 & -9.014e - 08 \\
-1.450e - 07 & 2.429e + 00 & -4.563e - 01 \\
-9.014e - 08 & -4.564e - 01 & 7.330e - 01
\end{pmatrix}.
$$

The underlying dataset is presented in [5] and combines 15581 data points and expert knowledge from an assembly line. Thus, we can ensure that the instrumental variable assumptions are satisfied.

---

[2] https://www.kaggle.com/datasets/veeralakrishna/aggressive-driving-data.

References

1. Cawley, J., Meyerhoefer, C.: The medical care costs of obesity: an instrumental variables approach. J. Health Econ. **31**(1), 219–230 (2012)
2. Cohn, D.A., Ghahramani, Z., Jordan, M.I.: Active learning with statistical models. J. Artif. Intell. Res. **4**, 129–145 (1996)
3. Drton, M., et al.: Algebraic problems in structural equation modeling. In: The 50th Anniversary of Gröbner Bases, vol. 77, pp. 35–87. Mathematical Society of Japan Tokyo (2018)
4. Ferguson, T.S.: A Course in Large Sample Theory. Routledge, London (2017)
5. Göbler, K., Windisch, T., Drton, M., Pychynski, T., Roth, M., Sonntag, S.: CausalAssembly: generating realistic production data for benchmarking causal discovery. In: Causal Learning and Reasoning, pp. 609–642. PMLR (2024)
6. Henckel, L., Perković, E., Maathuis, M.H.: Graphical criteria for efficient total effect estimation via adjustment in causal linear models. J. R. Stat. Soc. Ser. B Stat. Methodol. **84**(2), 579–599 (2022)
7. Isserlis, L.: On a formula for the product-moment coefficient of any order of a normal frequency distribution in any number of variables. Biometrika **12**(1/2), 134–139 (1918). http://www.jstor.org/stable/2331932
8. Johnson, A., et al.: MIMIC-IV. PhysioNet, pp. 49–55 (2020). https://physionetorg/content/mimiciv/1.0/. Accessed 23 Aug 2021
9. Kalisch, M., Bühlman, P.: Estimating high-dimensional directed acyclic graphs with the PC-algorithm. J. Mach. Learn. Res. **8**(3), 613–636 (2007)
10. Kilbertus, N., Kusner, M.J., Silva, R.: A class of algorithms for general instrumental variable models. Adv. Neural. Inf. Process. Syst. **33**, 20108–20119 (2020)
11. Murphy, S.A.: An experimental design for the development of adaptive treatment strategies. Stat. Med. **24**(10), 1455–1481 (2005)
12. Pallmann, P., et al.: Adaptive designs in clinical trials: why use them, and how to run and report them. BMC Med. **16**, 1–15 (2018)
13. Saar-Tsechansky, M., Provost, F.: Active sampling for class probability estimation and ranking. Mach. Learn. **54**, 153–178 (2004)
14. Staiger, D.O., Stock, J.H.: Instrumental variables regression with weak instruments (1994)
15. Taylor, M.C., Lynam, D., Baruya, A.: The effects of drivers' speed on the frequency of road accidents. Transport Research Laboratory Crowthorne (2000)
16. Yuan, M., Lin, Y.: Model selection and estimation in the gaussian graphical model. Biometrika **94**(1), 19–35 (2007)

Exploring the Use of Q-Learning in Causal Inference for Adaptive Interventions

Sha Zhou[1], YanHua Jiang[2], ZhiWei Jin[2], ZhenZhen Qian[3], MengMeng Ji[4], Chi Liu[5], HongYi Li[5], GuoWei Xuan[5], YuXing Shuai[6], and XinLin Chen[6(✉)]

[1] Second Affiliated Clinical School, Guangzhou University of Chinese Medicine, Guangzhou 510403, Guangdong, China
[2] Institute of Clinical Basic Medicine, China Academy of Chinese Medical Sciences, Beijing 100700, China
[3] Postdoctoral Station of China Academy of Chinese Medical Sciences, Beijing 100700, China
[4] Ningxia Medical University, YinchuanNingxia 750004, China
[5] Guangdong Provincial Hospital of Traditional Chinese Medicine, Guangzhou 510120, Guangdong, China
[6] Basic Medical College, Guangzhou University of Chinese Medicine, Guangzhou 510006, Guangdong, China
1019028995@qq.com

Abstract. Causal inference aids in estimating treatment effects for informed decision-making. Over time, there has been an increasing focus on customizing and tailoring intervention services, leading to the emergence of adaptive interventions. These interventions implement personalized sequences of treatment options by employing decision rules that integrate participant data to formulate customized recommendations for intervention progression. However, causal inference in adaptive interventions faces several challenges, a key one being the distinction between correlation and causation. This requires robust study designs, examples include the Sequential Multiple Assignment Randomized Trial (SMART), which is resource-intensive and complex. In addition, SMART data often contain confounding variables that mask the true interplay between interventions and their outcomes. Another obstacle is the diversity of populations; adaptive interventions that are effective in one subgroup may be ineffective in another. Response-adaptive designs potentially reducing the likelihood of selecting the most appropriate treatment for each individual.

In this context, we present Q-learning, an extension of regression analysis designed for situations where decisions are made sequentially regarding choices or intervention options available. Utilizing Q-learning involves applying it to assess the effectiveness of intervention options. Specifically, we utilize Q-learning alongside linear regression to approximate the optimal sequence of decision rules. This integration demonstrates how Q-learning, when combined with SMART data, enhances the refinement of decision rules, surpassing those originally embedded within the SMART framework. Eventually, we illustrate this methodology using the adaptive interventions for BPPV SMART (Reg. No. ChiCTR2100048603).

Keywords: Adaptive Intervention · Sequential Multiple Assignment Randomized Trials (SMART) · Q-Learning; Individualized Treatment · Causal Inference

X. Zhou and J. Jia (Eds.): PCIC 2024, CCIS 2200, pp. 86–94, 2025.
https://doi.org/10.1007/978-981-97-7812-6_8

1 Introduction

The acknowledged advantages of adaptive interventions [1] in the behavioral and social sciences are extensively documented, as highlighted in prior research. In adaptive interventions, the structure and/or the level of intensity are tailored specifically to individuals, considering their unique characteristics or clinical manifestations, and are further modulated in response to their evolving progress [2]. The implementation of the adaptive intervention concept can be achieved through the utilization of decision rules [3], which establish a connection between subjects' attributes and their ongoing progress, guiding the selection of specific subsequent intervention strategies, including the type and dosage/intensity. These customized intervention choices are determined by participants' values on variables that affect the effectiveness of specific intervention options, including both baseline characteristics and time-sensitive factors. This underscores the need to adapt the type or intensity of interventions based on these moderating variables.

Recently, researchers focusing on interventions have shown growing interest in experimental designs and analytical approaches. These approaches aim to identify optimal decision criteria for promoting high-quality adaptive interventions that are tailored to maximize the overall effectiveness of individualized intervention sequences [4–6]. The sequential multiple assignment randomized trials (SMART) [7] was developed to generate data tailored for the creation of adaptive interventions. Within SMART, participants move through various intervention stages, representing crucial decision points. At each stage, individuals are randomly assigned among different intervention options. Nahum-Shani et al. [8] offer techniques for comparing intervention choices across various phases of adaptive interventions and evaluating the comparatively straightforward adaptive interventions integrated into a SMART. Nevertheless, investigators frequently aim to develop interventions of greater complexity than those embedded in SMART. As an illustration, researchers frequently gather data on potential moderators, such as baseline individual and/or contextual characteristics, as well as adherence to and/or adverse reactions from prior interventions. Researchers intend to utilize this data to investigate the customization of intervention strategies according to these variables. Specifically, they are interested in leveraging the additional SMART data to refine the tailoring of adaptive interventions. Consequently, it is imperative to employ data analysis techniques that facilitate the development of the most optimal sequence of decision criteria, incorporating additional potentially beneficial variables for tailoring at each intervention stage.

Causal inference endeavors to estimate the impact of treatments on outcomes [9], a prevalent challenge in medical research. Typically, two types of studies are employed to assess treatment effects: randomized controlled trials (RCTs) [10, 11] and observational studies (OS) [12]. Randomized controlled trials are widely recognized for their rigorous methodology and capacity to reduce bias through random assignment. These trials randomly allocate participants into either a treatment group or a control group, representing the most effective approach to estimating treatment effects. However, traditional randomized controlled trial has been directed primarily at evaluating fixed interventions, non-adaptive interventions. Emerging as a novel paradigm for managing chronic relapsing diseases, adaptive interventions are gaining prominence in treatment and long-term care strategies. Causal inference in adapting interventions faces several challenges, one primary difficulty is distinguishing between correlation and causation. This necessitates rigorous study designs, such as SMART, which can be resource-intensive and complex. Additionally, SMART data often contain confounding variables that obscure the true relationship between an intervention and its outcomes. Another challenge is the heterogeneity of populations; interventions that work well in one subgroup may not be effective or could even be harmful in another. Response-adaptive designs might inadequately address diverse patient characteristics [13], such as genetic variability, potentially diminishing the likelihood of identifying the most suitable treatment for individual patients. These complexities make it essential for researchers to employ robust statistical methods and thoughtful study designs to draw meaningful causal conclusions and effectively adapt interventions.

In this section, we introduce Q-learning [14], a revolutionary and intuitive computational method for crafting more effective adaptive interventions based on existing data. Our Q-learning implementation employs a succession of linear regressions to formulate an ideal sequence of decision criteria. This sequence, essentially a series of adaptive intervention strategies, aims to maximize continuous outcomes. Q-learning evaluates each decision's efficacy—that is, the chosen intervention—at pivotal stages, incorporating the impact of preceding and subsequent adaptive choices.

2   Q-Learning: An Analytical Approach for Developing Adaptive Interventions

The core premise of constructing adaptive interventions lies in estimating potential outcomes, where patients are assigned to different treatment protocols that yield different values of response variables, indicating different degrees of efficacy. By comparing these potential outcomes, the optimal expected efficacy is identified and the corresponding treatment regimen represents the adaptive interventions at the current stage. However, these potential outcomes are not directly observable in practice, i.e., we cannot experience all possible efficacy outcomes for a given patient under different treatment protocols. In 2000, Lavory and Dawson presented a study that bridged the gap between potential outcomes and trial data using an adaptive intervention method called the SMART design [15]. This approach estimates potential outcomes based on SMART data, compares them to maximize expected effectiveness, and thereby constructs adaptive interventions. In the SMART design, patients are randomized at each decision juncture to ensure that treatment assignment remains independent of future potential outcomes. The trial framework guarantees that adaptive interventions can be formulated based on the data observed during the study. In 2012, Murphy published a study to comprehensively introduce SMART designs, including four SMART types, and Q-Learning modeling methods corresponding to different types of designs and methods for solving Adaptive Interventions [8].

Q-learning process involves an approximate dynamic planning method that initially estimates the conditional expectations (expected efficacy values) for future stages based on the patient's current medical history. This approach assumes optimal decisions are made at all subsequent decision points, and then solves for the adaptive interventions, which is called the Q function. The conditional expectation is called the Q-function. In Q-Learning and related methods, the Q function can be organized in a parametric, semiparametric, or nonparametric fashion [16].

3   Q-Learning: Two-Stage Reverse Regression

Our focus lies in utilizing SMART study data to formulate the ideal sequence of decision guidelines. Through estimation of Q-functions using Q-learning, we construct the afore-mentioned optimal sequence. Q-learning generally employs diverse regression tech-niques—such as linear, non-parametric, or additive regression—which can accommo-date various types of outcomes, including longitudinal, binary, ordinal, and continuous outcomes. For clarity, Q-learning [20] is implemented with linear regression for a contin-uous outcome (Y). In this case, the Q-function in the second stage might be formulated as

$$Q_2(O_1, A_1, Q_2, A_2; \gamma_2, \alpha_2)$$
$$= \gamma_{20} + \gamma_{21}O_1 + \gamma_{22}A_1 + \gamma_{23}O_1A_1 + \gamma_{24}O_2 + (\alpha_{21} + \alpha_{22}A_1 + \alpha_{23}O_2)A_2, \quad (1)$$

where, $\gamma_2 = (\gamma_{20}, \gamma_{21}, \gamma_{22}, \gamma_{23}, \gamma_{24})$, and $\alpha_2 = (\alpha_{21}, \alpha_{22}, \alpha_{23})$. Our primary focus is directed towards the parameters α_2, as they encapsulate crucial information regarding the second-stage intervention (A_2) should adjust according to the candidate tailoring variables (specifically A_1 and O_2). Referring to Eq. (1), it becomes evident that the optimal second-stage intervention choice (a_2) maximizes Q_2 is the one that maximizes the term $(\alpha_{21} + \alpha_{22}A_1 + \alpha_{23}O_2)a_2$. If $(\alpha_{21} + \alpha_{22}A_1 + \alpha_{23}O_2) > 0$, the term $(\alpha_{21} + \alpha_{22}A_1 + \alpha_{23}O_2)a_2$ attains its maximal value by $a_2 = 1$; if $(\alpha_{21} + \alpha_{22}A_1 + \alpha_{23}O_2) < 0$, the term $(\alpha_{21} + \alpha_{22}A_1 + \alpha_{23}O_2)a_2$ achieves its highest value through $a_2 = -1$. We conduct regression to estimate the vector parameters γ_2 and α_2 as follows:

$$Y \sim \gamma_{20} + \gamma_{21}O_1 + \gamma_{22}A_1 + \gamma_{23}O_1A_1 + \gamma_{24}O_2 + (\alpha_{21} + \alpha_{22}A_1 + \alpha_{23}O_2)A_2.$$

The next step is to evaluate the effectiveness of the optimal second stage selection. This includes

$$\tilde{Y}_i = \max_{a2} Q_2(O_{1i}, A_{1i}, O_{2i}, \alpha_2; \hat{\gamma}_2; \hat{\alpha}_2), i = 1, ..., n.$$

In this context, $\tilde{Y}_i$ simplifies to

$$\tilde{Y}_i = \hat{\gamma}_{20} + \hat{\gamma}_{21}O_{1i} + \hat{\gamma}_{22}A_{1i} + \hat{\gamma}_{23}O_{1i}A_{1i} + \hat{\gamma}_{24}O_{2i} + |\hat{\alpha}_{21} + \hat{\alpha}_{22}A_{1i} + \hat{\alpha}_{23}O_{2i}|. \quad (2)$$

$\tilde{Y}_i$ represents the anticipated mean result achieved when opting for the optimal intervention in the second stage, considering the available information

In our approach, linear regression is applied to the initial stage Q-function.

For the second-stage Q_2, the process involves constructing confidence intervals and conducting hypothesis testing on regression coefficients. It's important to highlight that this regression is standard in nature. Therefore, in extensive datasets, bootstrap meth-ods can be employed to gauge standard errors, establish confidence intervals, and per-form hypothesis tests. Evaluating estimators for regression coefficients in Q_1, however, presents less conventional inference challenges.

$$Q_1(O_1, A_1; \gamma_1, \alpha_1) = \gamma_{10} + \gamma_{11}O_1 + (\alpha_{11} + \alpha_{12}O_1)A_1, \quad (3)$$

where, $\gamma_1 = (\gamma_{10}, \gamma_{11})$, and $\alpha = (\alpha_{11}, \alpha_{12})$. Referring to Eq. (3), the primary choice in the initial stage (A_1) that maximizes Q_1 corresponds to the value of A_1 that optimizes the term $(\alpha_{11} + \alpha_{12}O_1)A_1$; that is, if $(\alpha_{11} + \alpha_{12}O_1) > 0$, $A_1 = 1$ maximizes the term $(\alpha_{11} + \alpha_{12}O_1)A_1$, and if$(\alpha_{11} + \alpha_{12}O_1) < 0$, $A_1 = -1$ maximizes the term $(\alpha_{11} + \alpha_{12}O_1)A_1$.

4   Using Q-learning to Analyze Data from BPPV SMART Study

4.1   BPPV SMART Study: Realization of Individualized Chinese Medicine Diagnosis and Treatment

As An example of SMART applied to Traditional Chinese Medicine (TCM) clinical trials, we explored the use of a multicenter, SMART two-phase design to carry out a clinical trial of Zexie Decoction in *Essentials of the Golden Chamber* for the discriminative treatment of phlegm-dampness-type BPPV [17] (Reg. No. ChiCTR2100048603) in an attempt to demonstrate the feasibility of SMART design for clinical trials in TCM. (see Fig. 1). In the first phase, a randomized controlled trial of Zexie Decoction (TZ) and Betahistine(C) was conducted, where enrolled patients were randomly divided into the control group (C) and the test group (TZ), and were treated with Betahistine and Zexie Decoction, respectively.

After 2 weeks of treatment, an evaluation of the therapeutic effect was conducted based on the Dizziness Handicap Inventory (DHI), which served as the primary indicator. In Stage 2, patients who had previously undergone Betahistine (C) treatment continued with Betahistine(C) if it had been effective; otherwise, they were switched to TZ. Patients previously treated with Zexie Decoction (TZ) continued with TZ if it was deemed effective; otherwise, they entered a randomized group and received either Lingguizhugan Decoction (TL) or an intensified dose of Zexie Decoction (Intensified TZ).

4.2   Data Analysis Procedure 1: Second-Stage Intervention Optimization

In BPPV SMART, the decision to re-randomize hinges on an interim result and previous treatment received. These factors were crucial in determining whether a participant should undergo re-randomization. This scenario pertained to data from a two-stage adaptive intervention SMART study, where initially participants were randomized to one of two intervention options: either receiving C_1 ($A_1 = -1$), or receiving TZ ($A_1 = 1$). In the second phase, only participants who did not respond initially to the TZ treatment were reassessed (with a 50% probability) for two potential alternatives: to change either to TL ($A_2 = -1$) or to Intensified TZ ($A_2 = 1$). Participants with effective phase I TZ treatment, phase I treatment with C (whether effective or not) were not re-randomized.

To implement Q-learning in this scenario, we started by developing the regression model specifically tailored for the second phase.

$$Q_2(O_1, Q_2, A_2; \gamma_2, \alpha_2) = \gamma_{20} + \gamma_{21}O_1 + \gamma_{22}O_2 + (\alpha_{21} + \alpha_{22}O_2)A_2, \qquad (4)$$

Based on the patient information at the beginning of the second stage ($O_{11}, O_{12}, O_{13}, O_{14}, O_{15}, A_1, O_{21}, O_{22}, O_{23}, O_{24}, O_{25}$), to find the optimal second-stage treatment measure (A_2) for the patients who were ineffective in the first-stage TZ. According to Eq. (4), $O_1 = O_{11} + O_{12} + O_{13} + O_{14} + O_{15}$; $O_2 = O_{21} + O_{22} + O_{23} + O_{24} + O_{25}$. Using only the data of patients with ineffective first-stage TZ, by regression Eq. (5), estimating the vector parameters γ_2 and α_2.

$$Y_2 \sim \gamma_{20} + \gamma_{21}O_1 + \gamma_{22}O_2 + (\alpha_{21} + \alpha_{22}O_2)A_2, \qquad (5)$$

The value of $(\alpha_{21} + \alpha_{22}O_2)$ was then obtained by the values of the regression coefficients that had been estimated (Mean $\pm$ SD). If $(\alpha_{21} + \alpha_{22}O_2) > 0$, the patient's second-stage treatment regimen was recommended to be Intensified TZ (A = 1); if $(\alpha_{21} + \alpha_{22}O_2) < 0$, the patient's second-stage treatment regimen was recommended to be TL (A = −1).

4.3 Data Analysis Procedure 2: Optimizing Initial Interventions through Backward Analysi

Worked backwards in time to determine the optimal initial intervention choice (A_1). Accordingly, Eq. (3) was used to model Q_1. According to the first-phase Q-learning model Q_1, the decision rule suggested that if $O_1 < 8$, the patient's first-phase intervention recommended TZ, and if $O_1 > 8$, the patient's first-phase intervention recommended C.

5 Discussion

The BPPV SMART study was based on the SMART two-stage randomized controlled trial of Zexie Decoction for the treatment of phlegm-dampness BPPV in the *"Essentials of the Golden Chamber"*. It cross-introduced Q-Learning two-stage inverse regression analysis to customize the optimal adaptive interventions for patients with phlegm-dampness BPPV, and analyzed the methodology of customizing the optimal adaptive interventions and the results of these interventions.

5.1 Reflecting Patients' Individualized Symptom Characteristics Through "Customized Variables"

In SMART, the customized variables are the basis for making decision rules, the selection of interventions is adjusted according to the values of the customized variables, and the adaptability of interventions is also based on the selection of customized variables. The customized variables selected for this study were the patient's response to treatment (i.e., the value of the decrease in the patient's DHI score at the end of the two weekends) versus the patient's first phase of the intervention.

There are several choices of customized variables. First, the customized variables were basic patient characteristics, including risky and protective characteristics. Second, patients' responses to previous interventions can also be used as customized variables, like BPPV SMART. Third, whether or not the patient has received prior treatment, or the type of treatment received, can also be used as an adjustment variable. That is, based on the patient's previous treatment history, the chosen intervention can be tailored. This allows for adaptation according to the treatments previously administered to the patient. For example, in BPPV SMART, the choice of the phase 1 interventions TZ and C also influenced the choice of the patient's phase 2 interventions, thus making the phase 1 interventions one of the customization variables.

5.2 Optimal Adaptive Interventions can Provide Scientific Guidance for the Selection of Individualized Treatment Plans Based on "Evidence-Based Treatment"

The ultimate goal of the Optimal Adaptation Intervention is to provide scientific evidence to guide the selection of the next phase of the intervention. Optimal adaptive interventions have two characteristics: initially, these interventions are personalized, taking into account the patient's unique characteristics and specific requirements. Secondly, they are dynamic and can evolve over time, adapting to the patient's newer characteristics and changing needs over time. Through a series of decision rules, relevant existing information about the patient's disease characteristics and response to treatment is continuously incorporated to guide the choice of adaptive interventions at each treatment stage. It mimics the process of "diagnosis and treatment" in Chinese medicine clinical practice, where clinicians select interventions based on key symptom profiles obtained during the consultation and modify the interventions according to the patient's changing symptomatology, with the aim of achieving good clinical outcomes. Therefore, the optimal adaptive interventions can provide scientific guidance for the selection of individualized treatment plans in the process of "diagnosis and treatment".

5.3 Strengths and Challenges of Q-Learning in Causal Inference of Adaptive Interventions

Randomized controlled trials (RCTs) are commonly used in causal inference to evaluate how treatments affect outcomes. In RCTs, participants are randomly allocated to either a treatment or control group, representing the most effective method for estimating treatment effects. However, in practice, traditional randomized controlled trials (RCTs) evaluate mostly fixed interventions, non-adaptive treatment regimens, which cannot meet the needs of evaluating adaptive interventions, making researchers more inclined to explore an individualized, adaptive clinical trial design, and Sequential Multiple Assignment Randomized Trials (SMARTs) were born out of necessity. And in the past decade or so, SMART has attracted increasing attention and is characterized as a multi-stage randomized trial design that determines the optimal adaptive intervention based on an individual's response to the initial intervention. Therefore, it has become a challenge to estimate the effect of adaptive interventions on outcomes in order to construct optimal decision rules.

In this interdisciplinary study, Q-learning two-stage inverse regression analysis was introduced to maximize the expectation of efficacy, to customize the best treatment decision, and to provide scientific evidence to guide the selection of the next stage of intervention. Since the true multivariate distributions of patient symptom characteristics O_1, O_2, and intervention efficacy Y are unknown, the most effective order of decision guidelines is constructed in conjunction with the use of SMART study data, which is denoted by $\{O_{1i}, A_{1i}, O_{2i}, A_{2i}, Y_i\}$, with $i = 1,..., N$, where N denotes the overall participant count in the study, and patients are treated in each of the two intervention are randomly allocated in each of the two phases, i.e., A_1 and A_2 are randomized. At the end of the second phase/end of the first phase, an efficacy prediction model is constructed to predict the efficacy of different interventions in the next treatment phase by using the patient's information, such as the patient's response to the previous treatments as well as the patient's current up-to-date characteristics, to maximize the efficacy expectation, and then provide the patient with the interventions that correspond to the optimal efficacy in order to customize the optimal treatment strategy.

The Q-learning algorithm offers several advantages compared to single regression approaches. Initially, Q-learning effectively manages the anticipated effectiveness of subsequent interventions while evaluating initial intervention effectiveness. Secondly, Q-learning predictions of effectiveness integrate both direct and indirect impacts of the initial intervention, essential for tailoring the optimal treatment approach. Lastly, Q-learning minimizes potential biases from personalized variables and critical effectiveness metrics. However, the application of Q-learning algorithms still faces some challenges. First, its direct use in observational studies may result in biased outcomes because there may be confounders predicting the probability of delivering intervenor measures A_1 or A_2, reflecting selection bias due to nonrandomized treatment. Second, the inference challenges posed by non-triviality should be considered when applying Q-learning. The presence of a absolute value function in the model leads to non-differentiability. Standard methods fail to consistently approximate the estimation of regression coefficients in Q_1 as a result, since the absolute value function lacks differentiability at zero. This study uses a soft-threshold operation (soft-threshold operation), which reduces bias [18], but improvements still need to be explored. Third, two customized variables were chosen for this study. However, many studies usually collect a large amount of information from which customized variables are selected, and the selection of customized variables is critical for the customization of optimal treatment strategies; therefore, more research efforts are needed to develop and explore methods of selecting customized variables. Despite encountering numerous challenges, studies have demonstrated that Q-learning, a straightforward regression-based approach, enables the customization of optimal treatment strategies using SMART data. While this study focused exclusively on

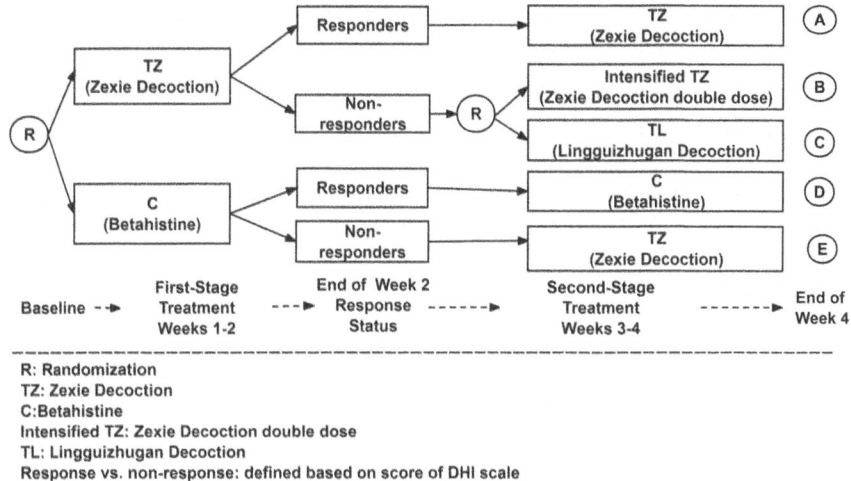

Fig. 1. A figure of BPPV SMART Study.

two intervention phases, Q-learning can extend to include studies with multiple phases beyond two.

References

1. Murphy, S.A., Lynch, K.G., Oslin, D., et al.: Developing adaptive interventions in substance abuse research. Drug Alcohol Depend. **88**(2), 24–30 (2007)
2. Dawson, R., Lavori, P.W.: Placebo-free designs for evaluating newmental health treatments: the use of adaptive interventions. Stat. Med. **23**(21), 3249–3262 (2004)
3. Kidwell, K.M., Postow, M.A., Panageas, K.S.: Sequential, multiple assignment, randomized trial designs in immuno-oncology research. Clin. Cancer Res. **24**(4), 730–736 (2018)
4. Brown, C.H., Ten Have, T.R., Jo, B., et al.: Adaptive designs for randomized trials in public health. Annu. Rev. Public Health **30**, 1–25 (2009)
5. Collins, L.M., Murphy, S.A., Strecher, V.: The multiphase optimization strategy (MOST) and the sequential multiple assignment randomized trial (SMART) new methods for more potent ehealth interventions. Am. J. Prev. Med. **32**(5), 112–118 (2007)
6. Rivera, D.E., Pew, M.D., Collins, L.M.: Using engineering control principles to inform the de-sign of adaptive interventions: a conceptual introduction. Drug and Alcohol Depend. **88**, 31–40 (2007)
7. Murphy, S.A.: An experimental design for the development of adaptive interventions. Stat. Med. **24**(10), 1455–1481 (2005)
8. Nahum-Shani, I., Qian, M., Almirall, D., et al.: Experimental design and primary data analysis for developing adaptive interventions. Psychol. Methods **17**(4), 457–477 (2012)
9. Pearl, J.: Causal inference in statistics: an overview. Statist. Surv. **3**(none) (2009). https://doi.org/10.1214/09-SS057
10. Colnet, B., Mayer, I., Chen, G.,et al.: Causal inference methods for combining randomized trials and observational studies: a review (2020)
11. Concato, J., Shah, N., Horwitz, R.I.: Randomized, controlled trials, observational studies, and the hierarchy of research designs. N. Engl. J. Med. **342**(25), 1887–1892 (2000)
12. Nichols, A.: Causal inference with observational data. Stata J.: Promot. Commun. Statist. Stata **7**(4), 507–541 (2007). https://doi.org/10.1177/1536867X0800700403
13. Xiong, W., Roy, J., Liu, H., et al.: Leveraging machine learning: covariate-adjusted Bayes-ian adaptive randomization and subgroup discovery in multi-arm survival trials. Contemp. Clin. Trials **142**(7), 107547 (2024)
14. Watkins, C.J.C.H.: Learning from delayed rewards. King's College Cambridge (1989)
15. Murphy, S.A., van der Laan, M.J., Robins, J.M.: CPPR Marginal mean models for dynamic regimes. JASA **96**, 1410–1423 (2001)
16. Zhao, Y., Kosorok, M.R., Zeng, D.: Reinforcement learning design for cancer clinical trials. Statist. Med. **28**(26), 3294–3315 (2009). https://doi.org/10.1002/sim.3720
17. Zhou, S.: An exploratory study on SMART design of Jin Gui Zexie Decoction (TZ) for the treatment of benign paroxysmal positional vertigo with phlegm-dampness. China Academy of Traditional Chinese Medicine, Beijing (2022)
18. Chakraborty, B., Murphy, S.A., Strecher, V.: Inference for non-regular parameters in optimal dynamic treatment regimes. Stat. Methods Med. Res. **19**(3), 317–343 (2010)

Author Index

B
Barahona, Mauricio   25

C
Chen, XinLin   86
Chung, Ha-Joon   49

D
Dhanakshirur, Mihir   25

F
Fang, Ying   41

H
Hong, Guanglei   49
Huang, Shanshan   65
Huang, Yen-Tsung   15

J
Ji, MengMeng   86
Jiang, YanHua   86
Jin, ZhiWei   86
Jun, Liao   65

L
Lai, En-Yu   15
Laumann, Felix   25
Li, HongYi   86

**Liu, Chi   86
Liu, Li   65
Luo, Tianjian   41

M
Mareis, Leopold   74

P
Park, Junhyung   25
Peng, Yun   1

Q
Qian, ZhenZhen   86

S
Shuai, YuXing   86
Sun, Huiyan   1

W
Wang, Kaijun   41
Wang, Lei   65

X
Xuan, GuoWei   86

Z
Zhao, Yonghe   1
Zhou, Sha   86

X. Zhou and J. Jia (Eds.): PCIC 2024, CCIS 2200, p. 95, 2025.
https://doi.org/10.1007/978-981-97-7812-6

GPSR Compliance

The European Union's (EU) General Product Safety Regulation (GPSR) is a set of rules that requires consumer products to be safe and our obligations to ensure this.

If you have any concerns about our products, you can contact us on ProductSafety@springernature.com

In case Publisher is established outside the EU, the EU authorized representative is:

Springer Nature Customer Service Center GmbH
Europaplatz 3
69115 Heidelberg, Germany

The manufacturer's authorised representative in the EU is Springer
Nature Customer Service Centre GmbH, Europaplatz 3, 69115 Heidelberg,
Germany. If you have any concerns regarding our products, please
contact ProductSafety@springernature.com

Printed and bound by CPI Group (UK) Ltd, Croydon, CR0 4YY
20/04/2026
02093294-0002